Piezoelectric Electromechanical Transducers for Underwater Sound

Boris S. Aronov

This book is subject to a Creative Commons Attribution-NonCommercial 4.0 International Public License (CC BY-NC 4.0). To view a copy of this license, visit https://creativecommons.org/licenses/ by-nc/4.0/. Other than as provided by these licenses, no part of this book may be reproduced, transmitted, or displayed by any electronic or mechanical means without permission from the publisher or as permitted by law.

Copyright © Boris Aronov, author, 2022

ISBN 9781644698211 (hardback)

ISBN 9781644698259 (Open Access)

Published by Academic Studies Press

1577 Beacon Street

Brookline, MA 02446, USA

press@academicstudiespress.com

www.academicstudiespress.com

*In memory of
my mentor, Lev Yakovlevich Gutin,
and my friends and colleagues, prominent Russian electroacousticians,
Lev Davidovich Lubavin and Vladimir Igorevich Pozern*

FOREWORD

The book is the most comprehensive coverage of piezoelectric acoustic transducers and all the related aspects of practical transducer designing for underwater applications in the field. It uses a physics-based energy method for analyzing transducer problems. This gives great physical insight into the understanding of the electromechanical devices. The great benefit of the energy method is that the multidisciplinary subject of electro-mechano-acoustics can be presented in parts and the solutions to the problems (electrical, electro-piezo, mechanical, and radiation) are combined using equivalent electrical circuit network theory. The energy and equivalent electromechanical circuit method at first is illustrated with transducer examples that can be modeled as a single degree of freedom system (such as spheres, short cylinders and flexural beams and plates). Then transducers are modeled as multiple degrees of freedom devices and the results are presented using multi contour electromechanical circuits. Special focus is made on the effects of coupled vibrations on the transducer performance. The Book gives also extensive coverage of acoustic radiation including acoustic interaction between the transducers. It provides practical results that are directly useful for the transducers modeling. While there have been many studies of acoustic radiation of various shapes, non-previous presented the results in terms of such practical utility.

    The book is inherently multidisciplinary. It provides essential background into vibration of elastic passive and piezoelectric bodies, piezoelectricity, acoustic radiation, and transducer characterization. Scientists and engineers working in the field of acoustics will find such a comprehensive treatment extremely useful not only for underwater acoustics, but also for electro-mechanics, energy conversion and medical ultrasonics.

David A. Brown, Dartmouth, Massachusetts, 2022

PREFACE

This book is initiated by the engineering experience of the author. Throughout his career the author has encountered many problems known to others involved in the design of electroacoustic transducers. The fact of the matter is that the complexity of designing electroacoustic transducers is inherent in the multidisciplinary nature of the subject. Therefore, the developers and designers of the transducers must possess the knowledge of several different theoretical disciplines (such as the vibration of mechanical systems, electromechanical conversion by deformed piezoelectric bodies, and acoustic radiation) and be able to actively use this knowledge to derive equations that describe the performance of the transducers. Furthermore, creating practical transducer designs that meet certain requirements and can operate under realistic environmental conditions requires the knowledge of properties of materials used and a certain level of engineering intuition that cannot be developed without a clear understanding of the underlying physics. Hardly anyone may possess all these capabilities without having received a specially targeted education, which, to the best of the author's knowledge, is not commonly available in the academic world. Usually, the necessary skills may be acquired through self-education, which was the case for the author. The main difficulties that arise in this endeavor are not in the lack of available information. On the contrary, the theoretical disciplines listed above are very well developed and are well-represented in the literature. Nevertheless, all these disciplines employ different methods for solving their problems and the results obtained are usually presented in forms not suitable for direct use in concert for synthesizing equations that govern transducer performance. Thus, the results must be tailored accordingly.

Experiencing the above difficulties over several decades, the author gradually developed a special approach to treating transducers problems that allows one to overcome many of the obstacles. The essence of this approach is in the consistent application of the physics-based energy method for solving all the problems that arise in the course of treating electromechanical and electroacoustic transducers. The first attempt to describe this concept was undertaken in *Electromechanical transducers from piezoelectric ceramic* published in 1990 in Russia. This version has now been updated and expanded to the extent that it can be considered a completely different book. Only the underlying energy approach to solving the problems has remained unaltered. This book is written for students, applied scientists and engineers in a way that should prove fruitful both for those who have only begun to chart their careers in electroacoustics as well as for those at a more advanced level. The content of the book is split into four p arts.

In Part I, titled "Introduction of energy method of treating the transducers," the main concepts of the method are considered (Chapter 1); applications of the method to calculating properties of transducers with single degree of freedom are illustrated (Chapter 2); and the study of problems for designing the transducers as a part of the transmit/receive channel is made (Chapter 3). The main concept is that of energy and following its transformation. Different types of energies involved in the electro-mechano-acoustic conversion in the course of transducer operation are presented in the generalized coordinates. All the governing equations are derived from the energy principles, that is, from the Law of Conservation of Energy for transducers with a single mechanical degree of freedom, and from the Principle of Least Action for transducers with multiple degrees of freedom. Equations describing the electromechanical part of the problem are reinterpreted as Kirchhoff's equations for the corresponding equivalent electromechanical circuits. In Chapter 2, the general approach is applied towards calculating the properties of transducers of widely used types (spheres, cylinders, bars undergoing extensional vibration and for circular plates and rectangular beams vibrating in flexure) that may be considered as systems with single mechanical degree of freedom. In Chapter 3, the operating properties of transducers as a part of a transmit/receive channel are considered and some recommendations regarding a rational transducer designing are presented. Given that the single degree of freedom approximation covers many practical transducer designs, Part I can be regarded as a self-sufficient study of underwater electroacoustic transduction on a basic level and can be read independently from the rest of the book.

The general treatment of electroacoustic transduction requires an advanced knowledge of the vibration of mechanical systems, electromechanical conversion in the deformed piezoceramic bodies and acoustic radiation. Information about these topics, which is necessary for the consideration of virtually all practical transducer types is presented in Chapters 4-6 of Part II under the title: "Subsystems of the Electroacoustic Transducers." All the constitutive equations are derived in these chapters from the Principle of Least Action as Euler's Equations in generalized coordinates. The obtained results are presented in the form of impedances, (including the radiation impedances), electromechanical transformation coefficients and acting forces (including those of acoustic origin) that can be directly substituted into the equivalent electromechanical circuits (multi contour in general) of the transducers. The diffraction coefficients and directional factors for differently configured transducer surfaces are also presented.

In Chapter 4, special attention is paid to the consideration of coupled vibrations in the generally two-dimensional mechanical systems. The results allow determining the range of aspect ratio, at which the system can be approximately considered as one-dimensional, where the problem can be simplified.

In Chapter 5, especial importance is ascribed to the theorem that sets the conditions, at which the electromechanical conversion under the longitudinal and transverse piezoelectric effects can be treated qualitatively in the same way. This allows for the unifying calculation technique for the transducers that employ these types of ceramics polarization. Another important subject is the general analysis of optimizing the effective coupling coefficients in nonuniformly deformed piezoceramic bodies.

Chapter 6 touches upon several noteworthy issues. Besides solving the general radiation problems, it provides a detailed consideration of the effects of baffling parts of the surfaces of cylindrical and spherical transducers, which ensures their unidirectionality. The technique for the experimental investigation of the acoustic interaction between transducers (or between the mechanically isolated parts of the same transducer) is also analyzed. Since the baffles have an effect on the acoustic near field, the interactions can rarely be treated analytically for practical transducer configurations, hence more reliable characterization of the interaction can be obtained through an experimental investigation.

The results obtained in the Part II are used in Part III of the book titled "Calculating transducers of different types" for synthesizing equations that describe the detailed operation of transducers of various configurations: cylindrical (Chapter 7), spherical (Chapter 8), plates and beams vibrating in flexure (Chapter 9) and bar transducers (Chapter 10).

Chapter 7 presents a study of cylindrical transducers that employ multimode extensional and flexural vibration of complete and incomplete cylinders (slotted cylinder projectors are also considered) for various practical applications. Different modes of cylinder polarization are considered, including the tangential polarization (with striped electrodes). An extensive study is provided of the effects of coupled vibrations on the electromechanical and acoustic performance of transducers that employ cylindrical piezoelements having finite thickness to diameter aspect ratios. Chapter 8 covers transducers which employ general multimode extensional vibrations of complete and incomplete piezoceramic spherical shells, (hemispherical in particular). The baffling of the parts of the surface that allows using multiple modes of vibration for unidirectional transducer operations is also considered.

In Chapter 9, a general analysis is provided of transducers which feature flexural vibrations of circular and rectangular piezoceramic plates (beams), including non-uniform over thickness and radius (length) transducer designs. Optimizing the effective coupling coefficients of the transducers is considered making use of the nonuniformity of the distribution of deformations in the volume of the plates. Corrections for transducer parameters due to a finite thickness to radius (length) ratio of the plates are taken into account. It is then concluded that the accuracy with which the wave numbers can be predicted substantially depends on the aspect ratio (especially for the higher modes of vibration) and presenting their values without the notion of the aspect ratios is not appropriate.

In Chapter 10, the length expander bar transducers are considered Transitions of configurations of bars to thickness vibrating plates at different polarizations and related dependencies of their effective coupling coefficients on the aspect ratios are considered using the technique of coupled vibrations. Relatively small attention is paid to the widely used Tonpilz transducer designs because they have already been described in detail in the available literature.

Part IV (Chapters 11 through 15) is titled: "Some aspects of the transducers designing."

In Chapter 11, a review of the existing data and some new results is presented regarding effects of operating environmental conditions, such as the hydrostatic pressure, temperature, and drive level on the parameters of piezoceramics. It is emphasized that, under these conditions, the parameters of ceramics may deviate significantly from those that are given in specifications for normal conditions. Moreover, they may differ for samples of ceramics supplied by different (and even by the same) manufacturers. This must be kept in mind when calculating the operating parameters of transducers under real conditions and in estimating a reasonable accuracy of calculation of the parameters. The variations in the parameters of transducers intended for operating at great depths can be avoided by using designs, which incorporate hydrostatic pressure compensation. Issues related to the practical implementation of the pressure compensation are examined in Chapter 12 (more general information), in Chapter 13 (regarding the liquid filled cylindrical projectors) and in Chapter 14 (regarding the hydrophones).

Chapter 13 presents some considerations regarding the practical challenges of the projectors design. Using the concept of the Reserves-of-Strength for improving parameters of the transducers of different types by optimizing their matching with the acoustic field is considered. The possibilities of increasing the dynamic and static mechanical strength of the projectors by prestressing and combining piezoceramic with passive materials in their mechanical systems are analyzed.

Chapter 14 is dedicated to the design of hydrophones and related issues. The hydrophones employing different transducer types are classified by the pressure and pressure-gradient hydrophones of the diffraction and motion types. Their properties as a source of energy of signal and internal noise for a receive channel are considered. Special attention is paid to the response of hydrophones and accelerometers to unwanted actions and to measures aimed at increasing their noise immunity.

Chapter 15 is crucial for the structure of the book because it introduces the practice of combining Finite Element Analysis (FEA) with analytical energy methods. This is illustrated with examples of flextensional and oval transducers. Combining powerful computer-based FEA techniques that are used to obtain results for vibration mode shapes with the energy method that yields great physical insight opens up a new area of research collaboration for many transducer problems. FEA allows the determination of the vibration mode shapes for mechanical systems that cannot be approximated analytically due to the complexities of the mechanical system and its boundary conditions.

The book also contains appendices with information on the properties of the piezoelectric ceramics and passive materials that may be used in transducer designs, and on the properties of the special functions that are referred to throughout the book.

In summary, the book presents methods for calculating the properties of most common electroacoustic transducer problems with particular focus on underwater applications. Moreover, by combining the FEA technique to determine the prerequisite vibration mode shapes with the energy method, virtually any transducer type may be analyzed. Still however, when it comes to choosing and designing a particular transducer for a particular application under demanding operating and environmental specifications – this remains somewhat of an art. Thus, recommendations of transduction choices for representative problems remain a guide and not a prescription for success.

It is inevitable that the book may contain typographical or content errors and thus the author would welcome the readers' comments and notifications of such.

Boris S. Aronov

ACKNOWLEDGEMENT

I am indebted to many people I have worked with throughout my career without whom this book could not have been written. Firstly, I am grateful to my mentor, the outstanding acoustician Lev Yakovlevich Gutin, who was my PhD advisor. He taught me that even complicated problems can be solved in a simple and physically clear way and demonstrated this throughout all his work. He was a strong proponent of using the energy physics-based approach to solving problems of electroacoustics, my first mentor in this regard.

It was my good fortune during the early part of my career to have worked for many years in the acoustic laboratory of Morphizpribor (Russian Marine Research Institution) in Saint Petersburg (then Leningrad) that was headed by R. E. Pasynkov and O.A. Kudasheva. A creative and friendly atmosphere existed in the laboratory, which resulted in many fruitful scientific research accomplishments. Discussions with my friends and colleagues L. D Lubavin, V.I. Pozern, M. D. Smaryshev, E. L. Shenderov, V. E. Glazanov and advising my colleagues and Doctoral PhD students G. K. Screbnev, N.M. Gribakina, L. B. Nikitin were priceless experiences that profoundly impacted this book. The style and content of this book have been influenced by my Doctor of Science Dissertation (1974) and my previous text *Electromechanical Transducers from Piezoelectric Ceramic* (1990) that were to a great extent based on the results of that period of my career.

The next stage of my professional career is in United States of America, where I started working with Dr. David A. Brown of BTech Acoustics LLC and the University of Massachusetts Dartmouth. David and I became and remain close friends for more than 20 years, and collaborated on many research projects, journal papers, and on co-advising graduate PhD and MS students in Electroacoustics. Hence, I am grateful to David Brown for providing me an unprecedented opportunity to teach and advise students at the University and work at the same time on the practical realization of my research and development transducer projects with BTech Acoustics. The results of this collaboration have been rewarding and have greatly contributed to the content of the book.

Furthermore, I would like to acknowledge many of our graduate students, who were eager to learn and assist in putting theory into practice often devoting long hours to experiments and computations. Specifically, Dr. Corey Bachand for transducer subsystem modeling and countless laboratory projects, coauthor of many journal publications; Sundar Regmi for participating in the research of coupled vibrations, Dr. Sairajan Saragapani for stripe-electrode tangential polarization research, Dr. Tetsuro Oishi for baffled multimode cylindrical transducers and their mutual impedances; Dr. Yan Xiang, who performed calculations regarding radiation of the baffled cylindrical and spherical transducers and FEA modeling, and many more students and engineering technicians too numerous to count including Gregory Bridge, Austin Souza, Zach Souza, Glenn Volkema, and others.

Further, our work would not have been successful had it not been the generous sup-port from the Office of Naval Research (ONR) and especially Jan Lindberg, Michael Traweek, Michael Wardlaw, Fletcher Blackmon, as well as the entrepreneurial activity of BTech Acoustics LLC, the SBIR innovative research program, the unique facilities funded by the Commonwealth of Massachusetts – Advanced Technology and Manufacturing Center (ATMC-UMass), Center for Innovation and Entrepreneurship (CIE) and the Underwater Acoustics Tank Test Facility at the Center for Marine Science and Technology (CMAST).

I am immensely grateful to Dr. Corey Bachand and to my Russian colleague, Aleksey Leznikov, for their truly gigantic work in preparing the manuscript of the Book for the publication that included all the formatting and numerous graphical works.

My special thanks go to my sons, Igor and Vitaly, for their moral and financial support throughout this years-long effort.

Part I

Introduction to Energy Method of Treating the Transducers

TABLE OF CONTENTS

TABLE OF CONTENTS 1
CHAPTER 1 3
Introduction 3
1.1   Block Diagram of a Transducer 3
1.2   Concepts of the Generalized Quantities 5
1.3   Actions on the Transducers 7
   1.3.1   Actions as Functions of Time 7
   1.3.2   Actions as Functions of the Spatial Coordinates 8
1.4   Forms of energy involved. 10
   1.4.1   Kinetic Energy 11
   1.4.2   Potential Energy 12
   1.4.3   Losses of energy 13
   1.4.4   Electromechanical Energy 15
   1.4.5   Acoustic Energy 18
1.5   Energy Flow and Sign Convention 21
   1.5.1   Directed Energy Flow 22
   1.5.2   Sign Convention 23
   1.5.3   Examples Illustrating Directional Energy Flow 26
1.6   Energy Approaches to Calculating the Transducers 28
   1.6.1   Balance of Energies in a Transducer having One Degree of Freedom, Single Contour Equivalent Circuit 29
      1.6.1.1 Transmit Mode 29
      1.6.1.2 Receive Mode 31
   1.6.2   Energy Approach to Calculating Transducer having Multiple Degrees of Freedom 33
      1.6.2.1 Least Action Variational Principle and Euler Equations 33
      1.6.2.2 Multi contour Equivalent Electromechanical Circuits 35
1.7   References 39
CHAPTER 2 40
Designing Transducers 40
2.1   One Degree of Freedom Transducers 40
2.2   Spherical Transducer 40
2.3   Cylindrical Transducers 47
   2.3.1   Acoustic Field of the Infinitely Long Cylindrical Transducer 50
   2.3.2   Acoustic Field of the Finite Height Cylindrical Transducer 53
2.4   Uniform Bar Transducers 55
   2.4.1   Effective Coupling Coefficient of a Transducer 59

2.5	Mass Loaded Bar Transducer	60
2.6	Flexural Type Transducers	65
	2.6.1 Rectangular Beam Transducer	65
	2.6.2 Cantilever Beam Transducer	68
	2.6.3 Circular Plate Transducer	71
	2.6.4 Acoustic Field Related Parameters of the Transducers of Flexural Type	75
2.7	References	80

CHAPTER 3 81

Transducer Performance Analysis 81

3.1	Operation in Transmit Mode	81
	3.1.1 Transducer Input Impedance and Tuning Conditions	82
	3.1.1.1 On the Tuning Conditions	92
	3.1.2 Effectiveness Factor of the Transmit Channel and Efficiency of a Projector	103
	3.1.2.1 Efficiency of a Projector	105
	3.1.2.2 Efficiency of a Projector over a Frequency Band	111
	3.1.3 Maximum Acoustic Power Radiated by a Transducer and Its Limitations	112
	3.1.3.1 The Optimal Acoustic Load and the Maximum Power Radiated	115
	3.1.3.2 Reserves of Strength Coefficients	118
	3.1.4 Frequency Response of a Projector	119
	3.1.5 Operation of a Projector in a Broad Frequency Band	123
	3.1.5.1 About the Bandwidth of a Projector	123
3.2	Transducers in the Receive Mode	129
	3.2.1 Transducer as a Member of Receive Channel	129
	3.2.2 Sensor as a Source of Energy for the Receive Channel	130
	3.2.3 Noise Property of a Receive Channel, Requirement for the Sensor Sensitivity	139
	3.2.3.1 Internal Noise of a Sensor	139
	3.2.3.2 Matching a Sensor with Preamplifier	140
	3.2.4 Response of the Sensors to Unwanted Actions	143
3.3	References	150

LIST OF SYMBOLS 151

INDEX 155

CHAPTER 1

INTRODUCTION

1.1 Block Diagram of a Transducer

The electromechanical piezoelectric ceramics transducers (further "piezoceramics transducers" or just "transducers") are employed as a part of the devices that are intended for converting electric energy into the mechanical (acoustic) energy (transmit mode of operation), or, conversely, for converting the mechanical (acoustic) energy into energy of electrical signals (receive mode of operation). Two closely linked problems can be distinguished in the theory and practice of transducers designing: calculating the transducer's output characteristics at specified input actions (the direct problem) and optimizing the transducer's design and efficiency factors at specified operating conditions including environmental conditions and conditions of their matching with mechanical (acoustic) loads and related electronics (the inverse problem). In terms of formulating the direct problem the multiple energy conversions that are performed by the devices that include electromechanical transducers can be conveniently illustrated for theoretical analysis by the block diagram presented in Figure 1.1.

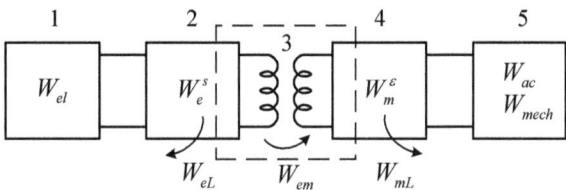

Figure 1.1: The block diagram of the electromechanical transducer. The blocks labels correspond to: (1) Electrical source of transducer (or load in case of receiver), (2) Electrical branch of the transducer, (3) Electrical to mechanical conversion, (4) Mechanical branch of the transducer; and (5) Mechanical (acoustic) load (or external source of mechanical energy).

In Figure 1.1, block 1 represents an external source of electric energy, or (depending on the direction of energy conversion) the first stage of processing the received signal; block 2 is the electrical part of a transducer (a unit to which electric energy is supplied from block 1, or as a result of the mechanoelectrical energy conversion); block 3 is a fictitious transformer that performs the electromechanical energy conversion; block 4 is the mechanical part of the transducer (a unit in which the vibration energy of the transducer system is concentrated being either produced as a result of the electromechanical energy conversion or supplied from an external source of mechanical energy); block 5 represents a consumer of mechanical energy (mechanical or acoustic load), or an external source of mechanical energy. The piezoelectric transducer as a system per se is represented by the blocks 2-4. Dividing the transducer system into subsystems possessing electric and mechanical energies is conditional, as the energies of these types are indivisible in the piezoelectric transducers. Dissipation of energy takes place in the process of the electromechanical transduction. This is accounted for in the block diagram by outgoing fluxes of energies of electric and mechanical losses.

The energies and energy fluxes (powers) involved in the transduction in the transmit mode of operation are denoted in Figure 1.1 as W and $\dot{W}$, respectively, with subscripts that correspond to the particular energy form. Thus, W_{el} is the total electric energy supplied to the transducer; W_e^S is the part of the electric energy stored on the electrical side of the "blocked" transducer (i.e., under the condition that the transducer cannot vibrate); W_{eL} is the energy of electrical losses; W_{em} is the part of supplied electrical energy, which is transformed into the mechanical energy of vibration of the transducer mechanical system at constant magnitude of electric field, W_m^E, by means of the mechanism of electromechanical conversion; W_{mL} is the energy of mechanical losses; W_{mech} and W_{ac} are the energies transferred into the mechanical or acoustic load.

Results of theoretical analysis in accordance with the block diagram may be applicable to whatever electromechanical transducers as far as characteristics of the mechanical load or external mechanical actions are known (the acoustical load and external action can be regarded as a particular case of the mechanical). In the case that electroacoustic transducers and specifically transducers for underwater applications are concerned, which is our intended goal, this analysis covers only a part of the direct problem. First, determining the acoustic load that is required for completing calculations of the transducer mechanical system vibration involves solving the problem of radiation by the transducer body (thus the overall problem becomes coupled mechanoacoustic). Finally, not only the acoustic power radiated by the transducer has to be determined, but also a spatial distribution of acoustic energy that is characterized by the directional factor of the transducer. Thus, in addition to the subsystems presented in the block diagram of Figure 1.1 the acoustic subsystem of a transducer must be considered.

The block diagram does not show the essential elements of the energy conversion system such as arrangements that are used for matching the electromechanical transducer with a source of excitation and a load. The matching elements may be incorporated in the source of excitation or in the load, but frequently they are a part of the transducer, and, consequently, they should be taken into account in the transducer analysis, especially, when considering issues of rational transducer designing (the reverse problem). It can be said without exaggeration that the optimal matching of an electromechanical transducer with a source of excitation and a load is one of the goals of its rational designing.

Due to reversibility of the piezoelectric transducers, we will predominantly consider one direction of energy conversion in the theoretical analysis, namely, the conversion of electrical energy into mechanical and acoustic energy, unless the peculiarities of the inverse conversion need to be emphasized.

1.2 Concepts of the Generalized Quantities

The main physical concept that will be used in our analysis is that of energy. The energy is measured by the work that has to be done in process of changing a physical system from an initial to final stage. Thus, expression for the mechanical work done by a constant force (f) that produces a displacement (ξ) of a body while acting in direction of the displacement is

$$W_{mech} = f\,\xi. \tag{1.1}$$

The energy (work) is measured in Joules (J), 1 J = N·m. The work done by a constant moment (torque) (m_τ) that turns a body at some angle (φ) is

$$W_\tau = m_\tau\,\varphi. \tag{1.2}$$

Amount of work done per unit of time is the energy flux or power ($\dot{W}$)

$$\dot{W} = \frac{dW}{dt}. \tag{1.3}$$

Thus,

$$\dot{W}_{mech} = f\,\dot{\xi} = f\,u, \quad \dot{W}_\tau = m_\tau\,\dot{\varphi} = m_\tau\,\varpi\,, \tag{1.4}$$

where $u = \dot{\xi}$ and $\varpi = \dot{\varphi}$ are the velocity and angular velocity, respectively. The power is measured in watts, 1 W = N·m/s.

At different stages of the process of energy conversion in the transducers the energies may be of different physical nature, but in all the cases the energy in general, W_g, can be represented as a product of two quantities, one of which can be considered as the generalized force, f_g, and another as the generalized coordinate (displacement), ξ_g, so that

$$W_g = f_g\,\xi_g\,. \tag{1.5}$$

The energy flux can be represented correspondingly as a product of the generalized force and the generalized velocity, u_g, namely,

$$\dot{W}_g = f_g\,u_g\,. \tag{1.6}$$

In general, when directions of the generalized force and generalized velocity do not coincide the energy flux is the scalar product (bold letters denote the vector quantities)

$$\dot{W}_g = \mathbf{f}_g \cdot \mathbf{u}_g\,. \tag{1.7}$$

In the case that the generalized parameters are not vector quantities by nature, the signs will be prescribed to them according to sign convention that will be discussed in Section 1.5.2.

Usually, the values of generalized coordinates and velocities may change with changes of system dimensions and of amount of substance in it (extensive values). The quantities of the generalized forces are independent of the dimensions and amount of substance in the system (intensive values). Thus, the mechanical force, moment of force, mechanical stress T, acoustic pressure p, and electric voltage v are the generalized forces; displacement ξ, turning angle φ, deformation S, and charge q are the generalized coordinates; velocity u, angular velocity $\tilde{\omega}$, and current i are the generalized velocities.

In general, a force may change in process of producing a work, and the expression for the work becomes

$$W_g = \int_I^{II} \mathbf{f}_g \cdot d\boldsymbol{\xi}_g\,, \tag{1.8}$$

where I and II are the initial and final position of a system, and both the generalized force and generalized displacement may be vectors.

1.3 Actions on the Transducers

Process of reduction of real actions on the electromechanical transducer that may have different physical nature to some generalized forces, and expression of results of these actions in the generalized coordinates or generalized velocities depends on the time and spatial characteristics of the actions. For unifying the theory of electromechanical transducers used for different applications, it is expedient to reduce the variety of actions to a combination of some standard functions, by means of which these actions can be easily expressed.

1.3.1 Actions as Functions of Time

It is appropriate to treat the piezoelectric ceramic transducers in the linear approximation unless the hard drive conditions are concerned. The related effects of nonlinearity will be considered in Chapter 11. Otherwise all the time depended actions may by expressed through the harmonic functions $\sin \omega t$, $\cos \omega t$ and, in the complex form, through the functions $e^{j\omega t} = \cos \omega t + j \sin \omega t$ by the general procedures applicable to the linear systems. It is noteworthy that using the negative exponential $e^{-i\omega t}$ is more traditional for description of the acoustic wave propagation, but the main problems of the electromechanical transducers designing are in the fields of mechanical and electric engineering, for which the positive exponential is common.

Thus, all the generalized forces and generalized coordinates (displacements) will be considered as the time harmonic functions. The instantaneous values of the generalized forces and generalized coordinates that were designated by small letters, will be represented by the corresponding capital letters in the complex form. For example,

$$f_g \to F_g = F_{og}(\omega)e^{j\omega t} \text{ and } u_g \to U_g = U_{og}(\omega)e^{j\omega t}, \tag{1.9}$$

where $F_{og}(\omega)$ and $U_{og}(\omega)$ are the complex amplitudes of the generalized force and velocity (only real values of the complex amplitudes correspond to the instantaneous values). From now on, the factor $e^{j\omega t}$ will be omitted for brevity and F_{og}, U_{og} will represent complex amplitudes of generalized forces and velocities.

When it comes to calculating the energies and powers in the complex form, we will introduce for them notations $\overline{W}$ and $\overline{\dot{W}}$, respectively. The complex power, for example, will be presented as product of the complex conjugated complex amplitudes of the generalized force and velocity, namely, as follows. We present the energy $\bar{W}_g$ and energy flux (power) $\dot{W}_g$ in the complex form as

$$\bar{W}_g = F_{og} U_{og}^* \qquad (1.10)$$

(the asterisk (*) denotes a complex conjugate quantity). In particular, $\bar{W}_{el} = UI^*$ is the complex value of the electrical power. The relationship between instantaneous and complex values of power (analogous for the energy), is

$$\bar{W}_g = \dot{W}_{gr} + j\dot{W}_{gx}, \qquad (1.11)$$

where

$$\dot{W}_{gr} = \frac{1}{T}\int_0^T \dot{W}_g dt \text{ and } \dot{W}_{gx} = \frac{2\pi}{T}\int_0^{T/4} \dot{W}_g dt - \frac{\pi}{2}\dot{W}_{gr} \qquad (1.12)$$

are the active and reactive power, respectively. It is obvious that any linear correlation between instantaneous values of power (or energy) is also valid for their active, reactive, and complex values.

The notion of a generalized complex impedance can be introduced in accordance with the relation

$$\bar{W}_g = \dot{W}_{gr} + j\dot{W}_{gx} = Z_g |U_{og}|^2. \qquad (1.13)$$

Taking into account formula (1.10), the generalized impedance can be represented as

$$Z_g = \frac{F_{og}}{U_{og}} = r_g + jx_g, \qquad (1.14)$$

where r_g and x_g are the active and reactive components of the complex impedance respectively.

1.3.2 Actions as Functions of the Spatial Coordinates

Actions as functions of the spatial coordinates can be applied at a point or distributed over the surface of a transducer. The mechanical system of a transducer can be often considered as having one degree of freedom. This means that the state of the system can be characterized by a

1.3. Actions on the Transducers

single variable. Thus, assuming that distribution of velocity of vibration in the mechanical system (mode shape of vibration) is known, the state of vibration of the mechanical system can be characterized by velocity of some point (the reference point), which can be called the reference velocity. Consider, for example, a simply supported bar under action of the concentrated force $F(x_1)$ and distributed along its length forces of density $\Delta F(x)$ that are shown in Figure 1.2. Let us assume that the mode shape of the vibrations, $\theta(x)$ of the bar is known, and $\theta(l/2) = 1$. Then the distribution of velocity along the bar is $U(x) = U_o \theta(x)$, where U_o is velocity of the reference point (reference velocity).

The energy flux supplied by the concentrated force $F(x_1)$ to the mechanical system, is

$$\overline{W} = F(x_1)U^*(x_1) = F(x_1)\theta(x_1)U_0^* = F_{eqv}U^*. \qquad (1.15)$$

Here $F_{eqv} = F(x_1)\theta(x_1)$ is the equivalent force, the action of which on displacement of the reference point produces the same work as the real force $F(x_1)$ produces on displacement of point x_1.

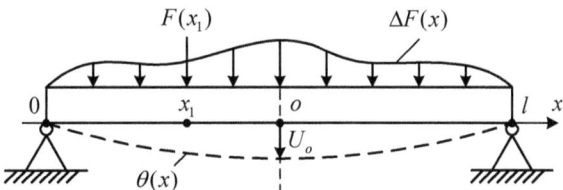

Figure 1.2: Simply supported beam under action of concentrated and distributed forces.

The energy flux supplied by the distributed force $\Delta F(x)$ to the mechanical system is

$$\Delta F(x)U^*(x)dx = U_0^*w\int_0^l \Delta F(x)\theta(x)dx, \qquad (1.16)$$

and in this case the equivalent force that produce the same work on displacement of the reference point as the real distribution of forces on the entire vibrating surface is

$$F_{eqv} = w\int_0^l \Delta F(x)\theta(x)dx. \qquad (1.17)$$

Since the values of the introduced equivalent forces depend on the position of the reference point, they can also be called reduced forces. The equivalent forces perform the same work on displacement of the reference point, as the actual forces do on the displacements of the transducer surface.

In general, the mechanical systems of transducers may have many degrees of freedom, and the nature of their vibrations is not known in advance. Suppose that some complete set of functions, $\{\theta_i(r)\}$, is determined for the surface of a mechanical system that meet the boundary conditions for the system (r is the radius vector of a point on the surface), then the unknown distribution of the vibration velocity over the surface, $U(r)$, can be represented by the expansion in terms of these functions:

$$U(r) = \sum_{i=1}^{\infty} U_i(r_o)\theta_i(r), \qquad (1.18)$$

where r_o is the radius vector of the point chosen as a reference point, $\theta_i(\bar{r}_o) = 1$.

Since the distribution of velocities and thus the state of a mechanical system becomes quite definite, when the coefficients $U_i(r_o)$ of the series are found, these coefficients can be adopted as the generalized velocities. When distributed forces act on the surface of a mechanical system, the equivalent force F_i that corresponds to the generalized velocity $U_i(r_o)$ can be determined as follows:

$$\bar{W} = \int_{\Sigma} \Delta F(r) U^*(r) d\Sigma = \sum_{i=1}^{\infty} U_i^*(r_o) \int_{\Sigma} \Delta F(r) \theta_i(r) d\Sigma, \qquad (1.19)$$

where from

$$F_i = \frac{\partial \bar{W}}{\partial U_i^*} = \int_{\Sigma} \Delta F(r)\theta_i(r) d\Sigma. \qquad (1.20)$$

It is convenient to choose the eigenfunctions of the problem for a particular transducer mechanical system vibration as the system of standard functions $\theta_n(r)$ that can also be called the supporting or coordinate functions. Thus, $\theta_n(x) = \sin(n\pi x/l)$ for a simply supported bar shown in the Figure 1.2, and the expression for the equivalent forces is

$$F_n = w\int_0^l \Delta F(x) \sin\frac{n\pi x}{l} dx. \qquad (1.21)$$

1.4 Forms of energy involved.

Consider expressions for the energies of different form that are involved in electroacoustic transduction with examples of systems having one degree of freedom (characterized by a single generalized coordinate).

1.4. Forms of energy involved.

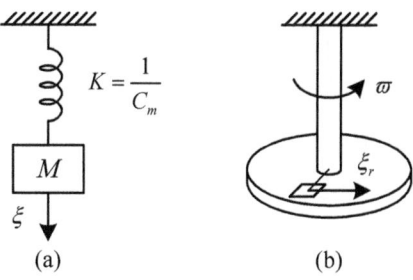

Figure 1.3: Examples of the mechanical systems with one degree of freedom.

1.4.1 Kinetic Energy

In the case that a force produces motion with velocity u of a body having mass M, in expression (1.7)

$$f = M\frac{d}{dt}u \text{ and } d\xi = udt, \qquad (1.22)$$

thus, the work done from the initial ($t = 0$) to the final state is

$$W_{kin} = M\int_0^t \frac{d}{dt}u \cdot u dt = \frac{1}{2}Mu^2. \qquad (1.23)$$

The work is stored in the energy of motion of the body, and can be released, when the body slows down. In the example shown in Figure 1.3 (a), the mass M can move linearly along the spring, and $u = \dot{\xi}$. In the case that a body having volume $\tilde{V}$ rotates, as is shown in Figure 1.3 (b), $u = r\varpi$, the force acting on an element of volume is $f = r\dot{\varpi}dm$, were dm is the mass of the element, and the kinetic energy of the body is

$$W_{kin} = \frac{1}{2}\varpi^2 \int_{\tilde{V}} r^2 dm = \frac{1}{2}I\dot{\varpi}^2, \qquad (1.24)$$

where

$$I = \int_{\tilde{V}} r^2 dm \qquad (1.25)$$

is the moment of inertia of the body.

In the case that a non-uniform distribution of velocity in the body takes place, such as $u(r) = u_o\theta(r)$, kinetic energy of the body will be determined by integrating the kinetic energies of elements of the body over its volume, thus,

$$W_{kin} = \frac{1}{2}\int_{\tilde{V}} \rho(r)u_o^2\theta^2(r)d\tilde{V} = \frac{1}{2}M_{eqv}u_o^2, \qquad (1.26)$$

where $\rho(r)$ is the density that in general can be non-uniform over the volume of the body, and M_{eqv} is the equivalent mass of the body

$$W_{kin} = M\int_0^t \frac{d}{dt}u \cdot u\, dt = \frac{1}{2}Mu^2. \qquad (1.27)$$

The energy W_L stored in the inductance L through which electric current i flows is also kinetic by nature. It is energy of motion of electric charges.

$$W_L = \frac{1}{2}Li^2 = \frac{1}{2}L\dot{q}^2. \qquad (1.28)$$

1.4.2 Potential Energy

The work done by a force is stored in the potential energy of a system in the case that the force is acting against inherent in the system restoring (reaction) forces that tend to keep the state of the system unchanged. The typical examples of such systems are the spring shown in Figure 1.3 (a), capacitor and the unit element of elastic body ($\Delta x = \Delta y = \Delta z = 1$) that is presented in Figure 1.4.

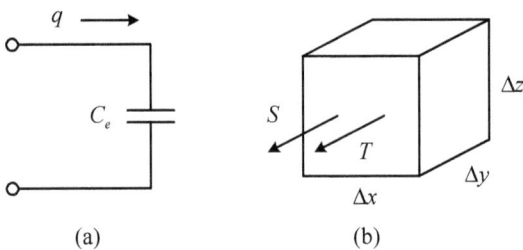

Figure 1.4: One degree of freedom systems possessing the potential energy.

The generalized reaction forces in these cases are: $f_{rp} = K\xi = \xi/C_m$ for the spring (subscript p stays for potential, K is the rigidity of the spring, and $C_m = 1/K$ is its compliance); $v = q/C_e$ for the capacitor (C_e is the capacitance); and $T = YS$ (TY is the Young's modulus). The generalized displacements are: displacement ξ, charge q and strain S, respectively. After performing integrations by formula (1.8) the expressions for the potential energies will be obtained as follows:

1.4. Forms of energy involved.

$$W_{mech} = \frac{1}{2}K\xi^2 = \frac{1}{2}\frac{\xi^2}{C_m}, \quad W_e = \frac{1}{2}\frac{q^2}{C_e} = \frac{1}{2}C_e v^2, \qquad (1.29)$$

$$w_m = \frac{1}{2}TS = \frac{1}{2}YS^2. \qquad (1.30)$$

The small letter in the expression for the elastic energy designates energy of the unit volume that must be integrated over the volume of a body to get the full potential energy. Given that $S = \xi/\Delta y$ and under the assumption that distribution of displacement over the body volume is $\xi(r) = \xi_o \theta(r)$, and the elastic properties can be in general not uniform the potential energy of the body will be determined as

$$W_{pot} = \frac{1}{2}\int_{\tilde{V}} Y(r)\xi_o^2 \theta^2(r) d\tilde{V} = \frac{1}{2}K_{eqv}\xi_o^2 = \frac{\xi_o^2}{2C_{eqv}}, \qquad (1.31)$$

where $Y(r)$ is distribution of the elastic modulus over the volume and K_{eqv} is the equivalent rigidity (C_{eqv} is the equivalent compliance) of the body

$$K_{eqv} = \frac{1}{2}\int_{\tilde{V}} Y(r)\theta^2(r)d\tilde{V}. \qquad (1.32)$$

The expressions for the potential energies are obtained under assumption that processes of changing state of the systems are linear. In this case all the stored energy will be given away when the systems return to their initial state.

1.4.3   Losses of energy

In reality some amount of energy converts into heat during deformation of the systems due to internal "friction" inherent in these processes. The term "friction" stays collectively for various mechanisms of energy loss that take place. Having negligible effect on the linearity of deformation, the losses of energy play significant role in balances of energy, especially when a system vibrates in the frequency range around its resonance.

The internal friction can be defined as the ratio

$$\frac{W_{gL}}{W_{g\,pot}} = \Delta_{gL}, \qquad (1.33)$$

where $W_{g\,pot}$ is the maximum generalized potential energy stored per cycle of deformation, W_{gL} is the part of the energy lost in this process, and Δ_{gL} may be called the specific loss coefficient[1]. The general assumption is that $\Delta_{gL} \ll 1$ in linear systems. The specific loss coefficients have to be determined through experimenting (as well as the elastic and dielectric constants of materials have to). The equivalent resisting force, f_{rL}, that is related to the energy of loss is assumed to be proportional to the generalized velocity (within limits of linearity of deformations), and it is directed as to oppose the deformation

$$f_{rL} = r_{gL}\dot{\xi}_g. \qquad (1.34)$$

Thus, formally the energy flux of losses can be presented as

$$\dot{W}_{gL} = f_{rL}\dot{\xi}_g = r_{gL}\dot{\xi}_g^2. \qquad (1.35)$$

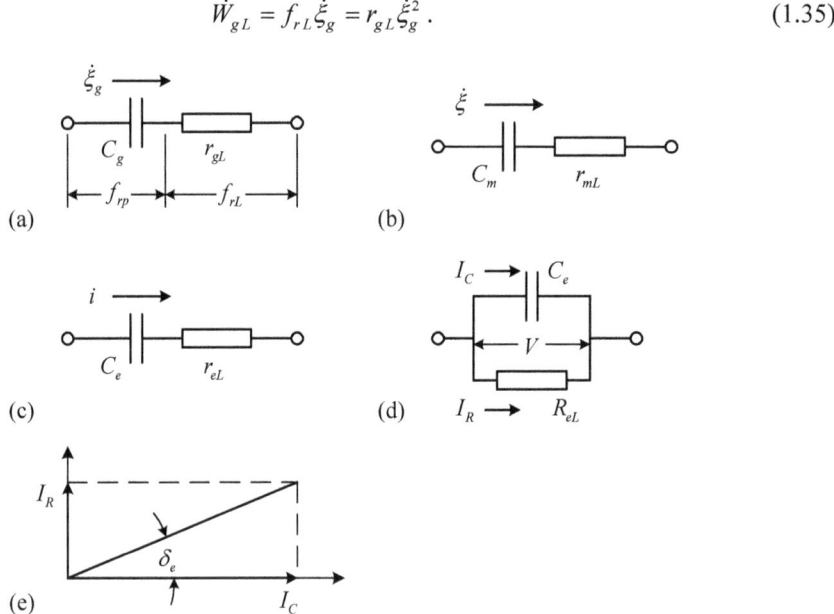

Figure 1.5: Equivalent representations of the resistances of mechanical and electrical losses.

The coefficient r_{gL} is the resistance of losses. In the case of mechanical energy $r_{gL} \to r_{mL}$ (resistance of the mechanical losses) and $\dot{\xi}_{gL} \to \dot{\xi}$, in the case of electrical energy $r_{gL} \to r_{eL}$ (resistance of dielectric losses) and $\dot{\xi}_{gL} \to i = \dot{q}$. If to represent the elements that possess the potential energy (generalized compliances C_g) and resistance of losses being in series connected, as shown in Figure 1.5 (a), then

$$\frac{W_{gL}}{W_{g\,pot}} = \frac{f_{rL}}{f_{rp}} = \frac{r_{gL}\dot{\xi}_g}{(\xi_g/C_g)} = r_{gL}\omega C_g = \Delta_{gL}. \qquad (1.36)$$

In the case of mechanical losses (Figure 1.5 (b))

1.4. Forms of energy involved.

$$\Delta_{mL} = r_{mL}\omega C_m = 1/Q_m, \tag{1.37}$$

where Q_m is called the mechanical quality factor.

In the case of dielectric losses (Figure 1.5 (c)) it is more common to present the capacitor and resistance of losses as connected in parallel. The equivalent parallel resistance, R_{eL}, will be found under the condition that $r_{eL} \ll (1/\omega C_e)$ that is equivalent to $\Delta_{gL} \ll 1$, as

$$R_{eL} \approx \frac{1}{r_{eL}(\omega C_e)^2}. \tag{1.38}$$

Considering the vector diagram of voltages vs. currents for the circuits in Figure 1.5 (c), will be obtained that

$$\Delta_{eL} = r_{eL}\omega C_{el} = \frac{1}{R_{eL}\omega C_{el}} = \tan\delta_e, \tag{1.39}$$

where δ_e is the phase angle between the active and reactive currents (voltages) that is called angle of dielectric losses.

The reason behind introducing different characterizations of mechanical (elastic) and electrical (dielectric) loss coefficients through the quality factor and angle of dielectric losses is due to different experimental methods for their determining. Thus, whenever the elements responsible for the potential energies of different nature are used, they must be accompanied by the corresponding resistances of losses. Expressions for the corresponding energy fluxes of mechanical and electric losses can be represented as

$$\dot{W}_{mL} = r_{mL}\dot{\xi}^2 = r_{mL}u^2, \quad \dot{W}_{eL} = \frac{v^2}{R_{eL}}. \tag{1.40}$$

1.4.4 Electromechanical Energy

The term electromechanical energy is introduced to characterize process of energy conversion that occurs in block 3 of the block diagram of Figure 1.1. Concept of the electromechanical energy (it is called "mutual" in Ref. 2) is fundamental for the theory of electromechanical conversion. Analysis of this concept under general assumptions regarding distribution of deformations in the piezoceramic bodies will be produced in Chapter 5. Here the concept of electromechanical energy will be illustrated with an example of a unit volume of piezoceramic material shown in Figure 1.6. The dimensions of the piezoelement are $\Delta x = \Delta y = \Delta z = 1$.

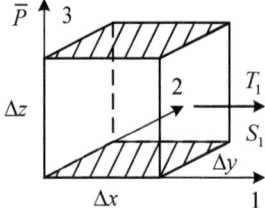

Figure 1.6: Piezoelement for illustrating concept of the electromechanical energy.

Status of the piezoelement that is polarized in direction of axis 3 and is mechanically loaded in direction of axis 1 while all its other surfaces are free of stress ($T_2 = T_3 = 0$) may be described by the piezoelectric equations[2]

$$S_1 = s_{11}^E T_1 + d_{31} E_3, \qquad (1.41)$$

$$D_3 = d_{31} T_1 + \varepsilon_{33}^T E_3, \qquad (1.42)$$

in which T_1 and E_3 are independent variables, (The standard notations of the piezoelectric material constants are used, as they are defined in Ref. 2). If to express T_1 from Eq. (1.41) and substitute the expression obtained into Eq. (1.42), the set of piezoelectric equations becomes

$$T_1 = \frac{1}{s_{11}^E} S_1 - \frac{d_{31}}{s_{11}^E} E_3, \qquad (1.43)$$

$$D_3 = \frac{d_{31}}{s_{11}^E} S_1 + \varepsilon_{33}^{S_1} E_3, \qquad (1.44)$$

where $\varepsilon_{33}^{S_1} = \varepsilon_{33}^T (1 - k_{31}^2)$ is the dielectric constant of the piezoelement "clamped" in direction of axis 1, $k_{31}^2 = d_{31}^2 / \varepsilon_{33}^T s_{11}^E$ is the electromechanical coupling coefficient square.

The internal energy of the piezoelectric element in the general case may be expressed as

$$w_{int} = \frac{1}{2} S_1 T_1 + \frac{1}{2} E_3 D_3 \qquad (1.45)$$

where $S_1 T_1 / 2$ and $E_3 D_3 / 2$ are the independent mechanical and electrical energies supplied by external sources; S_1, T_1 are the strain and stress along the axis 1 induced by an external source; E_3, D_3 are the electric field and charge density. If to consider the unit piezoelectric element as the energy converter in the "transmit mode", then the energy enters the piezoelement from the electrical side only, and the stress $T_1 = 0$ in Eq. (1.43) as independent variable. The electrical energy supplied to the piezoelement is

1.4. Forms of energy involved.

$$W_{el} = \int_0^V V dq = \int_0^{E_3} E_3 dD_3 \qquad (1.46)$$

(for the unit volume $V = E_3 \Delta z = E_3$ and $q = D_3 \Delta x \Delta y = D_3$). Using Eq. (1.44) we obtain expression for the total electric energy supplied to the piezoelement in the form of

$$w_{el} = \frac{1}{2}\frac{d_{31}}{s_{11}^E} S_1 E_3 + \frac{1}{2}\varepsilon_{33}^{S_1} E_3^2 = w_{em} + w_e^{S_1}. \qquad (1.47)$$

Here $w_e^{S_1} = \varepsilon_{33}^{S_1} E_3^2 / 2$ is the part of electrical energy supplied by the external source that is stored as electrical energy of piezoelement clamped along axis 1, and the term

$$w_{em} = \frac{1}{2}\frac{d_{31}}{s_{11}^E} S_1 E_3 \qquad (1.48)$$

will be defined as the density of electromechanical energy. It follows from Eq. (1.43) at $T_1 = 0$ that $d_{31} E_3 = S_1$. Upon substituting this expression into relation (1.48) will be obtained that

$$w_{em} = \frac{1}{2}\frac{d_{31}}{s_{11}^E} S_1 E_3 = \frac{1}{2}\frac{S_1^2}{s_{11}^E}. \qquad (1.49)$$

The right-hand term in this expression is the mechanical energy of a piezoelectric element determined with elastic constant s_{11}^E at constant electric field. It can be denoted as w_m^E. Note that $1/s_{11}^E$ is the analog of the Young's modulus of a passive elastic body and the expression (1.30) for the density of elastic energy must be replaced by

$$w_m^E = \frac{1}{2}\frac{S_1^2}{s_{11}^E}, \qquad (1.50)$$

when it is related to a piezoelectric element. Thus,

$$w_{em} = \frac{1}{2}\frac{d_{31}}{s_{11}^E} S_1 E_3 = w_m^E, \qquad (1.51)$$

and electromechanical energy can be considered as the part of electrical energy supplied to piezoelement that is transformed into the mechanical energy of deformation calculated with the elastic constant at constant electric field.

Formula (1.48) for the density of electromechanical energy can be generalized for a piezoelectric body of a finite size having one mechanical degree of freedom under the same mechanical boundary conditions $(T_2 = T_3 = 0)$. Given that $S_1 = \xi_1 / \Delta y$ and $E_3 = v / \Delta z$, where

$\Delta y = \Delta z = 1$, and that expression for the displacement in the body can be presented as $\xi = \xi_o \theta(r)$, it can be assumed that

$$W_{em} = \frac{1}{2}\frac{d_{31}}{s_{11}^E}\int_{\tilde{V}} \xi E_3 d\tilde{V} = \frac{1}{2}n\xi_o v, \qquad (1.52)$$

where $\tilde{V}$ is the volume of the body. The coefficient n will be called coefficient of electromechanical transformation. It must be determined as result of integrating a factual distribution of displacements and electric field in the body. Examples of calculating the coefficient of electromechanical transformation will be considered in Chapter 2. Expression for the electromechanical energy flux in the complex form, $\overline{W}_{em}$, will be found as

$$\overline{W}_{em} = \frac{1}{2}n(\dot{\xi}_o^* V + \xi_o^* \dot{V}) = nU_o^* V. \qquad (1.53)$$

Note that results of calculating the total potential energy by formula (1.31) and value of the equivalent rigidity (compliance) depend on the boundary electrical conditions, when they are related to a piezoceramic body, as it follows from expression (1.50) for density of potential energy. Therefore, the notations of the rigidity (compliance) must be marked with superscripts that show under what electrical conditions they were calculated. In most of the cases this is condition of constant electric field ($E = 0$). Therefore, the rigidities and compliances will be marked as K_{eqv}^E and C_{eqv}^E until otherwise will be noted.

1.4.5 Acoustic Energy

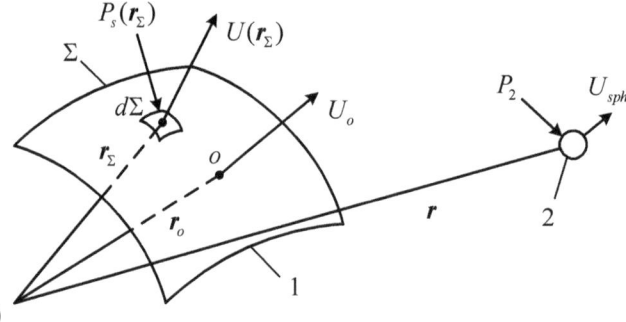

Figure 1.7: Illustration of the mechano-acoustic system consisting of the surface of radiating transducer 1 and pulsating sphere of a small radius 2.

1.4. Forms of energy involved.

Suppose that in the general case the acoustic field is generated by vibration of the transducer surface Σ with velocity distribution on it $U(r_\Sigma) = U(r_o)\theta(r_\Sigma)$ (Figure 1.7). In the figure r_Σ is the radius vector of the running point on the surface, and r_o is the radius vector of the reference point. Velocity at the reference point $U(r_o)$ will be further denoted for brevity as U_o. Expression for the acoustic energy flux (power) radiated by the transducer surface is

$$\overline{W}_{ac} = \int_\Sigma P_s(r_\Sigma) U_o^* \theta(r_\Sigma) d\Sigma, \qquad (1.54)$$

where $P_s(r_\Sigma)$ is the sound pressure acting on the surface of transducer in course of its vibration. Values of the sound pressure have to be determined by solving the radiation problem for a particular transducer surface configuration and mode of vibration.

Formulation of the radiation theory problems and some examples of determining the acoustic field related parameters of particular transducer types will be considered in Chapter 2. So far, we will assume that solution of the radiation problem is known, and the sound pressure produced by a transducer in the space can be represented as follows

$$P(r) = U_o \int_\Sigma P_\delta(r, r_\Sigma) \theta(r_\Sigma) d\Sigma = \frac{\rho c}{R} U_o \chi(r, \omega) e^{-j(kR - \frac{\pi}{2})}, \qquad (1.55)$$

where $P_\delta(r, r_\Sigma)$ is the sound pressure produced by an elementary source vibrating with unit volume velocity located on the otherwise clamped surface of the transducer; ρ, c are the density and sound speed in the medium; $k = \omega/c$, $\chi(r,\omega)$ is the function that characterizes sound pressure distribution in space. Function $\chi(r,\omega)$ is determined by the configuration and mode shape of vibration of the radiating surface. At the large distances R from the surface (at $R > \pi d^2 / \lambda$, where d is the maximum overall dimension of the radiating surface and λ is the acoustic wavelength), $\chi(r,\omega)$ (it will be called diffraction function) becomes independent of R. By substituting expression (1.55) for the sound pressure into formula (1.54) one obtains

$$\overline{W}_{ac} = \rho c |U_o|^2 \int_\Sigma \frac{\chi(r_\Sigma, \omega)}{r_\Sigma} e^{-j(kr_\Sigma - \pi/2)} \theta(r_\Sigma) d\Sigma = Z_{ac} |U_o|^2. \qquad (1.56)$$

where from the reaction of acoustic field on the transducer vibration–acoustic radiation impedance Z_{ac} may be determined after the radiation problem is solved. Thus,

$$\overline{W}_{ac} = Z_{ac} |U_o|^2 = (r_{ac} + jx_{ac}) |U_o|^2, \qquad (1.57)$$

where r_{ac}, x_{ac} are the active and reactive components of the radiation impedance.

In the receive mode the external acoustic field presents the mechanical energy source for a transducer. Determining parameters of this source constitutes a problem that is formulated as follows. If $P(r_\Sigma)$ is the acoustic pressure on the surface of the transducer under the action of acoustic field, and $U(r_\Sigma) = U_o \theta(r_\Sigma)$ is the distribution of vibrations generated by this action, then the mechanical energy flux supplied by the acoustic field to the transducer (we will call this energy acoustomechanical, W_{am}) will be

$$\overline{W}_{am} = \int_\Sigma P(r_\Sigma) U_o^* \theta(r_\Sigma) d\Sigma. \tag{1.58}$$

The sound pressure $P(r_\Sigma)$ on the vibrating surface of a transducer may be represented as

$$P(r_\Sigma) = P_u(r_\Sigma) - P_s(r_\Sigma), \tag{1.59}$$

where $P_u(r_\Sigma)$ is the sound pressure that would be acting in acoustic field on the clamped surface of a transducer (at $U(r_\Sigma) = 0$) and $P_s(r_\Sigma)$ is the sound pressure generated on the surface of the transducer as a back radiation due to its vibration with velocity distribution $U(r_\Sigma)$. Taking into account expression (1.59), the relation (1.58) can be rewritten as follows

$$\overline{W}_{am} = \int_\Sigma P_u(r_\Sigma) U_o^* \theta(r_\Sigma) d\Sigma - \int_\Sigma P_s(r_\Sigma) U_o^* \theta(r_\Sigma) d\Sigma. \tag{1.60}$$

The second term in this expression is $\overline{W}_{ac} = Z_{ac} |U_o|^2$, as it follows from Eqs. (1.54) and (1.57). The first term may be represented as $F_{eqv} U_o^*$, where the designation is introduced

$$\int_\Sigma P_u(r_\Sigma) \theta(r_\Sigma) d\Sigma = F_{eqv}. \tag{1.61}$$

For actual calculating the equivalent forces the problem of diffraction of acoustic field on the clamped transducer surface must be solved that will result in determining sound pressure $P_u(r_\Sigma)$. These issues are considered in Chapter 2 for some transducer types and in Chapter 6 in the general formulation.

The flux of the acoustomechanical energy consumed by the transducer that is converted into the energy of vibration of its mechanical system can be expressed as

$$\overline{W}_{am} = \overline{W}_m = F_m U_o^* = Z_{inm} |U_o|^2. \tag{1.62}$$

Here F_m is the equivalent force that can be imagined as acting on the transducer surface at the reference point, and Z_{inm} is the input impedance of the mechanical system of the transducer reduced to this point. Finally, the relation (1.60) can be presented after summarizing expressions for the involved energies, as

$$F_m = Z_{inm} U_0 = F_{eqv} - Z_{ac} U_0. \qquad (1.63)$$

This relation can be illustrated by the equivalent mechanical circuit shown in Figure 1.8.

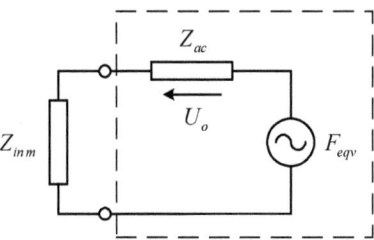

Figure 1.8: Equivalent acoustomechanical generator.

Thus, the acoustic field can be considered in respect to a transducer as an energy source with mechanical motive force F_{eqv} and internal impedance Z_{ac}. The part of energy of this source $Z_{inm} |U_0|^2$ is transmitted into the mechanical system of the transducer and another part $Z_{ac} |U_0|^2$ is reflected from the surface of the transducer (is consumed by the internal impedance of the source). Peculiarity of this source of energy is that both F_{eqv} and Z_{ac} depend on the configuration of radiating surface of the transducer and on its mode of vibration. The final expression for the acoustic energy supplied to mechanical system of a transducer may be presented as

$$\bar{\bar{W}}_{am} = (F_{eqv} - Z_{ac} U_0) U_o^* \qquad (1.64)$$

after combining expressions (1.62) and (1.63).

1.5 Energy Flow and Sign Convention

In the process of electroacoustic (and in the reversed acoustoelectric) transduction directional flow of energies of different physical nature takes place. With the energies being described in the generalized quantities the processes of energy propagation cannot be attributed to the particular geometrical coordinate systems, and considering the energy balances require introducing some signs convention regarding the direction of energy flow (positive or negative), and regarding the signs of the related to these energies generalized coordinates (displacements) and forces.

1.5.1 Directed Energy Flow

Consider the energy state of the volume $\tilde{V}$ bounded by the surface Σ (Figure 1.9).

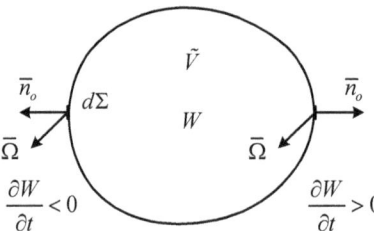

Figure 1.9: Closed volume for illustrating the directional energy fluxes.

Amount of energy inside the volume can change due to a spatial flow of energy through the surface. The rate of change of amount of energy inside the volume, W_g, should be equal to the total sum of energy fluxes through the boundary surface per unit time according to the energy conservation law. Direction of the spatial energy flow will be considered positive if the flux increases amount of energy inside the volume. Mathematical formulation of this statement was given by Russian physicist N. A. Umov.[3] He introduced a concept of the density vector of the energy flux of a physical field known as Umov's vector. We will denote Umov's vector $\bar{\Omega}$. By definition (in our notations)

$$\bar{\Omega} = w_g \boldsymbol{n}_u, \qquad (1.65)$$

where w_g is the density of the generalized energy and is the unit vector heading in direction of the generalized velocity. For example, for the flux of acoustic energy

$$\bar{\Omega} = p\bar{v}. \qquad (1.66)$$

The theorem formulated by N. A. Umov states that

$$\frac{\partial W_g}{\partial t} = -\int_\Sigma \bar{\Omega} d\bar{\Sigma} = -\int_\Sigma \bar{\Omega} \boldsymbol{n}_\Sigma d\Sigma, \qquad (1.67)$$

where $d\bar{\Sigma} = \boldsymbol{n}_\Sigma d\Sigma$ is the directional unit surface area, and $\boldsymbol{n}_\Sigma$ is the unit vector of the outward normal to the surface. On the other hand

$$\frac{\partial W_g}{\partial t} = \int_{\tilde{V}} \dot{w}_g d\tilde{V}. \tag{1.68}$$

According to the Gauss theorem applied to the vector field of vector $\bar{\Omega}$

$$\int_{\tilde{V}} \mathrm{div}\bar{\Omega} d\tilde{V} = \int_{\Sigma} \bar{\Omega} \boldsymbol{n}_\Sigma d\Sigma, \tag{1.69}$$

where the flux of vector $\bar{\Omega}$ is directed outside of the volume. Comparing expressions (1.68), (1.67) and (1.69) we conclude that $\dot{w}_g = -\mathrm{div}\bar{\Omega}$. This relation reflects the fact that in our case the convention is accepted that incoming flux of vector $\bar{\Omega}$ is positive. Using the expressions (1.65) and (1.67) we arrive at relation

$$\frac{\partial W_g}{\partial t} = -\int_{\Sigma} \bar{\Omega} \boldsymbol{n}_\Sigma d\Sigma = -\int_{\Sigma} \dot{w}_g \boldsymbol{n}_u \boldsymbol{n}_\Sigma d\Sigma, \tag{1.70}$$

where from the condition follows for the energy flux being positive

$$\dot{w}_g \boldsymbol{n}_u \boldsymbol{n}_\Sigma < 0. \tag{1.71}$$

At the same time the condition (see (1.7)) must be fulfilled

$$\dot{w}_g = \boldsymbol{f}_g \cdot \boldsymbol{u}_g > 0. \tag{1.72}$$

The energy flow is also positive if both inequalities are reversed. In the case that one of these inequalities is not fulfilled, the energy flux is negative (energy flows out of the system). Given that expressions for the energy flux densities $\dot{w}_g$ are known from Section 1.4, positive directions of the generalized velocities (of vector $\boldsymbol{n}_u$) have to be chosen for applying the concept of directional energy flow. This can be done by establishing a certain sign convention.

1.5.2 Sign Convention

Defining the signs for generalized displacements and velocities having different physical nature can be arbitrary, as long as these definitions are used consistently. Once the positive direction of a generalized velocity is chosen, the positive direction of corresponding generalized force must result in the positive direction of energy flow in accordance with inequality (1.72).

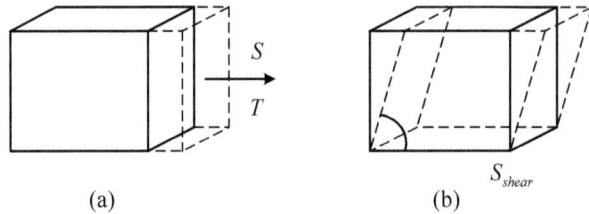

(a)   (b)

Figure 1.10: Illustration of the sign convention for the tensile and shear stress.

We accept the following rule of sign common for theory of elasticity and for mechanical and electrical engineering. For the acoustic field related quantities the rarefication will be considered as positive and compression – as negative. The tensile strain will be considered as positive. The shear deformations will be considered positive, if they result in decreasing the angle between coordinate axes compared with their initial state. Accordingly, the mechanical tensile stresses are positive ($TS > 0$) and the shear stresses that form a couple in the clockwise direction (Figure 1.10). Under the accepted rule of signs displacement of the mechanical system surface directed towards the external normal and the forces directed likewise or under sharp angle to the normal must be considered as conventionally positive, as illustrated in Figure 1.11 with example of a bar.

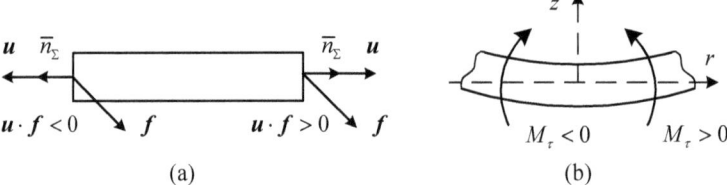

(a)   (b)

Figure 1.11: Illustration of the sign convention for displacement and bending moment.

The rule of signs for bending moments is illustrated in Figure 1.11 (b) with example of a beam under flexure. If the bending is produced with curvature convex downwards (which is positive), the moment is positive in the clockwise direction to the left of a cross section of the beam and in the counterclockwise direction to the right of a cross section. If the bending is produced with curvature convex upwards, the sign of curvature and the positive moments will change direction.

Increase of charge density inside a volume and direction of the electric intensity that yields a charge increase are also considered to be positive. According to the standard for the electrical engineering passive sign convention (PSC), the direction of current into the positive terminal of a component of a system and the voltage vector across its input that is pointed to the terminal are positive. As a result, the energy flow enters this part of the system, which thus presents a load for an external energy source (active element of the system). This situation is illustrated with Figure 1.12, where the source of electrical energy is also shown. Suppose that the referenced positive direction of the current is indicated by the unit vector n_i. If to formally assume that the outer normal to an element of the system is directed outside of its terminal along the wire, as it is shown in the figure by unit vectors n_Σ, then on the input of passive element $n_i \cdot n_\Sigma < 0$.

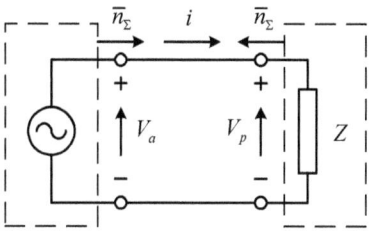

Figure 1.12: The sign convention for direction of current.

As the energy flux through this input $\dot{w}_{el} = i \cdot v_p > 0$ by the sign convention, the condition (1.71) is met, and the energy flows into the element. On the output of the active element $n_i \cdot n_\Sigma > 0$. The current that is flowing out of the positive terminal must be considered as negative. According to the second Kirchhoff's rule $v_a + v_p = 0$, and $v_a = -v_p$. Thus $\dot{w}_{el} = i \cdot v_a < 0$ and the sign of inequality (1.71) is reversed. This means that the energy flows out of the active element. In this situation the element, out of which energy flows, is a source of energy for the element, in which energy enters and which can be considered as a load.

The sign convention and considerations regarding direction of energy flow used in this example can be generalized to different energy forms, if to replace current with a generalized velocity and voltage with a generalized force.

The sign convention applies to alternating processes. Being defined for some moment of time (for example, for the first half of a cycle), the direction of energy flow does not change because both the generalized velocities and forces reverse directions simultaneously in the course of vibration.

Consider examples that illustrate directional flow of energy.

1.5.3 Examples Illustrating Directional Energy Flow

1. Longitudinal deformation of a bar by the forces acting on its ends (Figure 1.13).

In the case shown in Figure 1.13 (a) the bar is in the phase of extension ($u_0 > 0$, $u_1 > 0$), and forces acting on the ends are tensile ($f_0 > 0$, $f_1 > 0$). The energy flows enter the bar through both ends according to the condition (1.71). In the case (b) the bar experiences compression on the both ends ($u_0 < 0$, $u_1 < 0$), and the forces are tensile on the left end ($f_0 > 0$) and compressive on the right ($f_1 < 0$). According to the condition (1.71) the energy enters the bar through the right end and flows away through the left end. The general rule is that change of sign of either strain or stress reverses direction of the energy flow, whereas simultaneous change of signs of both quantities does not change direction of flow.

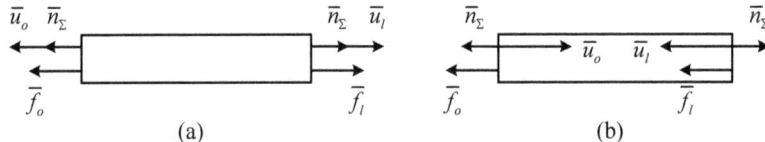

Figure 1.13: Directional energy flow of the longitudinal deformation of a bar.

2. "Horse and carriage" example (Figure 1.14).

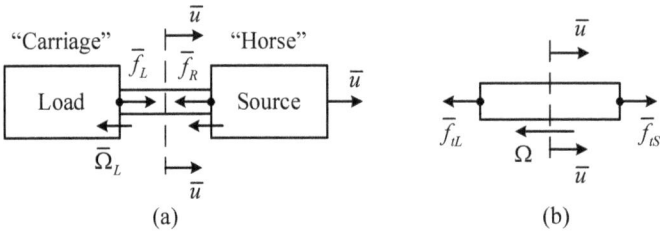

Figure 1.14: Energy flow in the "horse and carriage" situation.

Consider the system of two bodies connected mechanically with a rigid truss (or rod), in which one of the bodies tows another with velocity u. The acting force f_L is applied to the "carriage", and to the "horse" is applied the reaction force f_R. According to Newton's third law $f_R = -f_L$. Under action of these forces the truss experiences tension on the both ends, as shown in Figure 1.12 (b), because the reactive forces acting on the ends of the truss are equal and opposite to the forces f_R and f_L, correspondingly. By the condition (1.71) the energy flows out of "horse" through the truss and into the "carriage". This is shown by direction of vector $\bar{\Omega}$ in Figure 1.14. Thus, the "horse", as a body that produces outgoing energy flow, is the source of mechanical energy, and the "carriage" that consumes this energy is the mechanical load.

3. Electromechanical energy flow (Figure 1.15)

Expression for the electromechanical energy flux is, $\dot{W}_{em} = v u_g n$, where v is the voltage on the electric part of a transducer u_g is the generalized velocity of vibration on the mechanical part and n is the electromechanical transformation coefficient. Dimension of this quantity is such that the term vn has the dimension of force, $f_{em} = vn$, and the term $u_g n$ has the dimension of current, $u_g n = i$. Thus, $\dot{W}_{em}$ can be considered either as $\dot{W}_{em} = v \cdot i$ on the electric part, or as $\dot{W}_{em} = f_{em} \cdot v_g$ on the mechanical part. With conventionally positive directions of i, u, f_{em}, and v_g ($\dot{\xi}$), as shown in Figure 1.15, the flux of the electromechanical energy flows out of electric unit 2 and is directed into mechanical unit 4. Thus, the block 2 is the source of electrical energy and block 4 is the mechanical load.

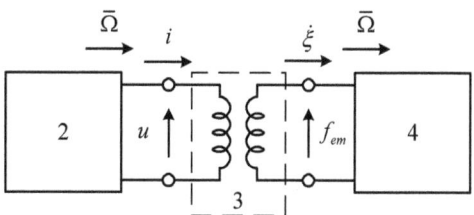

Figure 1.15: Illustration of the flow of electromechanical energy.

In the reversible systems the division of elements of a system in the sources and the loads is conditional. Source in one situation may become a load in another situation depending on direction of energy flow through the element. A typical example in this respect is the mechanical system of a transducer (block 4 in the block diagram of Figure 1.1). In the transmit mode it is a load for the electric energy source, and a source of energy for the acoustic field. In the receive mode block 4 is the mechanical load for source of the acoustic energy and the source of electric energy for block 2. In the equivalent circuits that will be used for describing balance of energies in the systems with multiple energy transformations the loads may be represented by the generalized impedances determined as $Z_g = F_g / U_g$, where the quantities of the generalized force and velocity depend on the form of energy. The two terminal block that represents generalized source of energy shown in Figure 1.16 (a) can be replaced following Thevenin's theorem by the equivalent generalized generator with voltage E_{ocg} and internal impedance Z_{ing}, shown in Figure 1.16 (b). According the Thevenin's theorem the value of voltage equals to the open circuit voltage at the terminals, and the impedance is the input impedance of the block measured at terminals under condition that all voltage sources inside the block are short circuited and all current sources are open circuited.

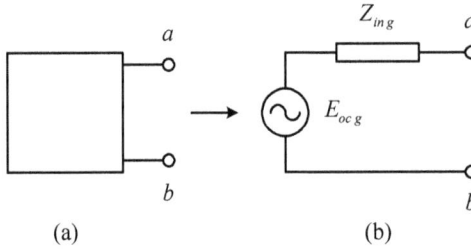

Figure 1.16: Generalized equivalent generator.

The terminology typical for the electrical circuits is used for denoting the generalized quantities, loads and sources of energy (voltage, current, generator, impedance) and for their graphical representation by several reasons. First, the equivalency can be established between parameters that determine expressions for energies of different physical nature through their generalized variables and those for electric energy. Second, and the most important is that calculating technique for the electric networks is well developed and can be applied to calculating characteristics of different systems after their operation is described in terms of equivalent electric circuits.

1.6 Energy Approaches to Calculating the Transducers

Considering energy status of an energy converting system, a balance can be made of changing its total energy ($W_{kin} + W_{pot}$) and incoming and outgoing energy flows based on the energy conservation law or on its equivalents. According to the convention accepted all the incoming flows must be regarded as positive (with signs (+)) and all the outgoing flows as negative (with signs (-)). Thus, due to the energy conservation law

$$\frac{d}{dt}(W_{g\,kin} + W_{g\,pot}) = \sum \dot{W}_{g\,in} - \sum \dot{W}_{g\,out}. \qquad (1.73)$$

The incoming energy flows can be represented as produced by the sources of energy – generalized equivalent generators. The outgoing flows can be considered as consumed by the loads and represented as the generalized impedances. Processes of vibration are considered to be adiabatic, i.e., occurring without exchange of thermal energy with surroundings. At first we will llustrate using the energy balances with examples of the mechanical (block 4 in Figure 1.1) and electrical (block 2) subsystems of a transducer, assuming that they have one degree of freedom each, i.e., the energy state of the subsystem depends on a single variable – displacement of the reference point ξ_o in the mechanical case and charge q in the electrical.

1.6.1 Balance of Energies in a Transducer having One Degree of Freedom, Single Contour Equivalent Circuit

1.6.1.1 Transmit Mode

For the mechanical block the incoming is electromechanical energy flux, $\dot{W}_{em}$, and the outgoing are the acoustic energy flux, $\dot{W}_{ac}$, that is generated by vibration of the mechanical system surface, and flux of energy of the mechanical loss, $\dot{W}_{mL}$. After converting Eq. (1.73) into the complex form and substituting expressions (1.26), (1.31), (1.34), (1.52) and (1.57) for the energies and energy fluxes involved we arrive at the equation

$$\left(j\omega M_{eqv} + \frac{1}{j\omega C_{eqv}^E} + r_{mL} + Z_{ac} \right) U_o = Vn. \qquad (1.74)$$

The expression in parentheses represents the mechanical impedance

$$j\omega M_{eqv} + \frac{1}{j\omega C_{eqv}^E} + r_{mL} + Z_{ac} = Z_m^E, \qquad (1.75)$$

and Eq. (1.74) may be rewritten as

$$Z_m^E U_o = Vn. \qquad (1.76)$$

For the electrical block incoming is the energy of an external electrical source, $\overline{\dot{W}}_{el} = VI^*$, and outgoing are the electromechanical energy that flows into the mechanical block and energy of the electrical loss. The potential energy is stored in the capacitance. Kinetic energy is absent so far as no inductances are involved (the special case of using inductances for tuning the input impedance of a transducer will be considered separately). After substituting expressions (1.29), (1.40) and (1.52) for the energies and converting to the complex form Eq. (1.73) becomes

$$I = \left(j\omega C_e^U + \frac{1}{R_{eL}} \right) V + nU_o . \qquad (1.77)$$

Here the first term represents the current that flows through the parallel connection of the capacitance and resistance of electrical losses, or it would be the total input current if the mechanical system of the transducer was clamped ($U_o = 0$). Therefore, the capacitance is marked with superscript U to underline that it is the input capacitance of mechanically clamped transducer. The second term can be transformed following Eq. (1.76) to the form

$$nU_o = \frac{n^2}{Z_m} V \qquad (1.78)$$

that makes it clear that it represents the current flowing through the impedance introduced into the electrical side of the transducer as result of transformation of the input impedance of the mechanical block. Thus, finally we arrive at the equation

$$I = \left(j\omega C_e^U + \frac{1}{R_{eL}} + \frac{n^2}{Z_m} \right) V , \qquad (1.79)$$

where from the input impedance of the transducer as a load for the external source of electrical energy will be found as

$$Z_{in\ tr} = \frac{V}{I} = \left(j\omega C_e^U + \frac{1}{R_{eL}} + \frac{n^2}{Z_m} \right)^{-1} . \qquad (1.80)$$

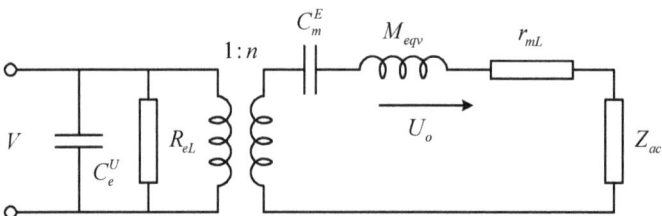

Figure 1.17: Equivalent electromechanical circuit of one degree of freedom transducer operating in the transmit mode.

The equations (1.74) and (1.79) can be obtained from the electrical circuit that is shown in Figure 1.17 under the condition that the electrical elements in the circuit have meaning of the equivalent mechanical parameters of a transducer. All the results of calculating produced with using this circuit are equivalent to those that can be obtained with the original equations. Therefore, the circuit is called the equivalent electromechanical circuit of transducer (or just equivalent circuit for brevity).

1.6.1.2 Receive Mode

In the receive mode the transducer is a source of electric energy regarding the input of a system of signal processing. Therefore, it can be represented by the equivalent generator, as shown in Figure 1.16, which requires for its characterization knowing the electromotive force (voltage across the open circuited output of transducer) and impedance of the transducer determined between the output terminals.

For the mechanical block the incoming is the acoustic energy flux, $\dot{W}_{am}$ by formula (1.64) and the outgoing are the flux of energy of the mechanical loss, $\dot{W}_{mL}$, and electromechanical energy flux, $\dot{W}_{em}$, that is produced by vibration of the mechanical system and by reaction of the electrical load that the open circuited electrical block presents that will be denoted V_{oc}. Formula (1.53) should read $\overline{\dot{W}}_{me} = nU_o^* V_{oc}$ (in this case the energy is called mechanoelectrical due to the opposite direction of flow). As the result, Eq. (1.73) being converted to the complex form will be

$$\left(j\omega M_{eqv} + \frac{1}{j\omega C_{eqv}^E} + r_{mL} + Z_{ac} \right) U_o + nV_{oc} = F_{eqv}. \quad (1.81)$$

For the electrical block incoming is the mechanoelectrical energy. The energy of the electrical loss is the only outgoing flow, due to absence of an electric load. The potential energy is stored in the capacitance. The resulting equation is

$$V_{oc} \left(j\omega C_e^U + \frac{1}{R_{eL}} \right) = nU_o. \quad (1.82)$$

Due to relation $(1/R_{eL}) \ll \omega C_e^U$, it will be found that

$$V_{oc} \approx \frac{n}{j\omega C_e^U} U_o, \quad (1.83)$$

and after substituting this expression for V_{oc} into Eq. (1.81) we finally obtain that

$$\left(j\omega M_{eqv} + \frac{1}{j\omega C_{eqv}^E} + \frac{n^2}{j\omega C_e^U} + r_{mL} + Z_{ac} \right) U_o = F_{eqv}. \quad (1.84)$$

The expression in parentheses represents the mechanoelectrical impedance

$$Z_{me}^E = j\omega M_{eqv} + \frac{1}{j\omega C_{eqv}^E} + \frac{n^2}{j\omega C_e^U} + r_{mL} + Z_{ac}, \quad (1.85)$$

and Eq. (1.84) may be rewritten as

$$Z_{me}^E U_o = F_{eqv}. \quad (1.86)$$

The Eqs. (1.84) and (1.82) follow from the electrical circuit in Figure 1.18 with elements having meaning of the equivalent mechanical and electrical parameters of the equations. This circuit is the equivalent circuit of a transducer in the receive mode. Similarity of the equivalent circuits of a transducer in the transmit and receive modes reflects the reciprocity of the electromechanical conversion performed by a piezoelectric transducer.

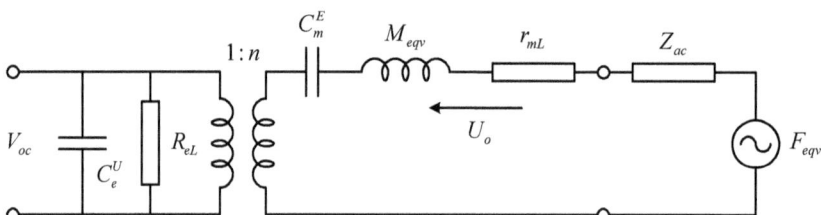

Figure 1.18: Equivalent electromechanical circuit of one degree of freedom transducer operating in the receive mode.

All the necessary calculations regarding operating characteristics of one degree of freedom transducers as loads and sources of energy can be made using the equivalent circuits of Figure 1.17 and Figure 1.18. Determining of the acoustic field related properties of the transducers require considering on the separate issues. This will be done in Chapter 6.

The one degree of freedom approximation for the transducer subsystems is appropriate for wide variety of transducers and may cover the most of their practical applications. Although in general, the mechanical system of an electroacoustic transducer must be considered as the system with multiple degrees of freedom. In other words, the displacement distribution in the mechanical system may be represented as a function of several independent variables – generalized coordinates. Thus, for example, if distribution of displacements in the mechanical system is represented in the form

$$\xi(r) = \sum_{i=1}^{N} \xi_i(r_o)\theta_i(r), \tag{1.87}$$

where $\{\theta_i(r)\}$ is a set of linearly independent functions satisfying the boundary conditions for the mechanical system, then the quantities ξ_i can be considered as the generalized coordinates for the system. In this case all the energies involved will be expressed through the generalized coordinates ξ_i and the generalized velocities $\dot{\xi}_i$. The electrical side of a transducer also may have several degrees of freedom (several electrical inputs), though they always have lumped parameters, because dimensions of the electromechanical transducers for underwater applications are much smaller than the electromagnetic wave lengths.

For treating transducers having multiple degrees of freedom using the energy conservation law is not sufficient, and different energy methods must be used. The most general and the most powerful of them is the Least Action Principle.

1.6.2 Energy Approach to Calculating Transducer having Multiple Degrees of Freedom

1.6.2.1 Least Action Variational Principle and Euler Equations

Equations of motion for a physical system can be derived in the general form (i.e., irrespective of a specific coordinate system) from the variational Least Action Principle.[4] According to this principle the equations of motion of the system may be found from the following condition of minimizing the function represented by the integral of the Lagrangian L, of the system taken between the fixed initial (at moment t_1) and final (at moment t_2) states of the system

$$\delta \int_{t_1}^{t_2} L(\xi_1, \xi_2, \ldots; \dot{\xi}_1, \dot{\xi}_2, \ldots) dt = 0. \tag{1.88}$$

The Lagrangian is a function of the generalized coordinates ξ_i and velocities $\dot{\xi}_i$. This function must be determined for a particular system. The equations of motion may be obtained from the condition (1.88) by means of calculus of variations and they are known as the Euler equations[4]

$$\frac{d}{dt}\left(\frac{\partial L}{\partial \dot{\xi}_i}\right) - \frac{\partial L}{\partial \xi_i} = 0 \quad (i = 1, 2, ...). \tag{1.89}$$

In general, the form of the Euler equations depends on the type of function that describes the state of the system. Thus, if the Lagrangian is a function $L_x(\xi, \dot{\xi}, \xi'_x)$ from the displacement of points, $\xi(x,t)$, its time derivative, $\dot{\xi}$, and the derivative in respect to coordinates, ξ'_x, the Euler equations, which in this case is derived from the condition

$$\delta \int_{x_1}^{x_2} \int_{t_1}^{t_2} L_x(\xi, \dot{\xi}, \xi'_x) dx dt = 0 \tag{1.90}$$

has the form

$$\frac{d}{dt}\left(\frac{\partial L_x}{\partial \dot{\xi}}\right) + \frac{\partial}{\partial x}\left(\frac{\partial L_x}{\partial \xi'_x}\right) - \frac{\partial L_x}{\partial \xi} = 0. \tag{1.91}$$

This equation can be generalized to the case of several spatial coordinates (then the term $\partial/\partial x (\partial L_x / \partial \xi'_x)$ becomes $\sum_i \partial/\partial x_i (\partial L_x / \partial \xi'_{x_i})$), or to the case of several functions $\xi_i(x,t)$. The latter case involves the system of equations of type of Eq. (1.74) for each of $\xi_i(x,t)$ functions. In order to apply the corresponding Euler equations to a specific system, it is necessary to determine the Lagrangian L or other function (of L_x type) that characterizes the state of the system, and to express this function in an explicit form in the system of adopted generalized coordinates. This task is of primary importance in applying the method under consideration to different systems (mechanical, electrical, electromechanical). In principle Lagrangian is a function that being used in the Euler equations leads to correct results of solving the particular problems, as those that agree with existing experience. Thus, for a conservative system the Lagrangian is used in the form

$$L = W_{kin} - W_{pot}. \tag{1.92}$$

This results in the particular case of the Euler equations that are commonly called the Lagrange equations

$$\frac{d}{dt}\left(\frac{\partial W_{kin}}{\partial \dot{\xi}_i}\right) + \frac{\partial W_{pot}}{\partial \xi_i} = 0, \quad i = 1, 2, \ldots. \tag{1.93}$$

They describe vibrations of a system in the generalized coordinates.

1.6. Energy Approaches to Calculating the Transducers

Based on the energy balance expressed by Eq. (1.73), we suggest the Lagrangian for electroacoustic transducer as a system with multiple energy conversions in the form

$$L = W_{g\,kin} - W_{g\,pot} + \sum W_{g\,in} - \sum W_{g\,out}. \tag{1.94}$$

It will be seen that using this function throughout this treatment never failed in getting correct equations of motion for different modifications of the transduction systems.

In the case of a non-conservative system that is supplied with external energy W_e under action of the generalized forces $f_i = (\partial W_e / \partial \xi_i)$, the Lagrange equations are

$$\frac{d}{dt}\left(\frac{\partial W_{kin}}{\partial \dot{\xi}_i}\right) + \frac{\partial W_{pot}}{\partial \xi_i} = \frac{\partial W_e}{\partial \xi_i} = f_i, \quad i = 1, 2, \ldots \tag{1.95}$$

If a volume element of an elastic body is considered as the system, then

$$L_x = w_{kin} - w_{pot} + w_e, \tag{1.96}$$

where w_{kin}, w_{pot} are the kinetic and potential energies of the volume element; w_e is the energy supplied to the volume element by external action. The energies w_{kin}, w_{pot}, w_e depend on the geometric coordinates. In the most cases w_{kin} depends only on $\dot{\xi}$, w_{pot} on ξ'_x, and w_e on ξ; thus the Euler equations in the form (1.91) are applicable. After substituting the expression (1.96) for L_x into Eq. (1.91) the differential equation of motion of the body in geometrical coordinates will be obtained as

$$\frac{d}{dt}\left(\frac{\partial w_{kin}}{\partial \dot{\xi}}\right) - \frac{\partial}{\partial x}\left(\frac{\partial w_{pot}}{\partial \xi'_x}\right) = \frac{\partial w_e}{\partial \xi}. \tag{1.97}$$

At $(\partial w_e / \partial \xi) = 0$ this equation describes free vibration of the body.

Thus, we see that all the problems of vibration of both transducer as a whole and particular mechanical systems employed in the transducers can be solved by using the energy based approach to deriving equations of motion regardless of coordinate systems, which is convenient for a particular case – generalized or geometrical.

1.6.2.2 Multi contour Equivalent Electromechanical Circuits

In order to obtain equations that describe vibration of the mechanical system of a transducer out of the general Euler equation (1.89), all the energies involved into expression (1.94) for the

Lagrangian must be determined. How it is done can be demonstrated with an example of the kinetic energy. After substituting expression (1.87) for $\xi(r)$ into the general formula for the kinetic energy

$$W_{kin} = \frac{1}{2}\int_{\tilde{V}} \rho(r)\dot{\xi}^2(r)d\tilde{V} \tag{1.98}$$

will be obtained

$$W_{kin} = \frac{1}{2}\int_{\tilde{V}} \rho\left(\sum_{i=1}^{N} \dot{\xi}_i \theta_i(r)\right)^2 d\tilde{V} = \frac{1}{2}\sum_{i=1}^{N}\sum_{l=1}^{N} M_{il}\dot{\xi}_i\dot{\xi}_l, \quad i,l = 1, 2, \ldots, N, \tag{1.99}$$

where $M_{il} = (1/\dot{\xi}_i)(\partial W_{kin}/\partial \dot{\xi}_l)$ can be considered as the equivalent masses. In the case that functions $\theta_i(r)$ are orthogonal (for example, if they form a system of eigenfunctions for the problem of mechanical system vibration), the mutual terms with $i \neq l$ disappear, and

$$W_{kin} = \frac{1}{2}\sum_{i}^{N} M_{ii}\dot{\xi}_i^2, \quad i = 1, 2, \ldots, N. \tag{1.100}$$

Expressions for all the other energies in formula (1.94) for the Lagrangian can be represented in the analogous way. This will be done in course of considering particular types of the mechanical systems. After substituting the Lagrangian into Eq. (1.93) and transferring the result to the complex form, we arrive at the set of equations that describe vibration of the transducer mechanical system in the transmit mode,

$$\left(j\omega M_{eqvi} + \frac{1}{j\omega C_{eqv\,i}^E} + r_{mLi} + Z_{aci}\right)U_{oi} = Vn_i, \quad i = 1, 2, \ldots, N. \tag{1.101}$$

Here the equivalent parameters with number i are attributed to the generalized velocity $\dot{\xi}_i = U_{oi}$. Due to assumed orthogonality of functions θ_i the equations are independent.

Equation for the electrical subsystem may be obtained from the energy balance (1.73) in the same way as Eq. (1.77) was obtained, but in this case expression (1.53) for $\overline{W}_{em}$ must be replaced by

$$\overline{W}_{em} = V^* \sum_{i=1}^{N} n_i U_{oi}, \tag{1.102}$$

where n_i is the coefficient of electromechanical transformation determined for the distribution of strain according to function θ_i. Thus, the equation will be

1.6. Energy Approaches to Calculating the Transducers

$$I = \left(j\omega C_e^U + \frac{1}{R_{eL}} \right) V + \sum_{i=1}^{N} n_i U_{oi} . \quad (1.103)$$

The set of Equations (1.101) and (1.103) is equivalent to the multi contour electrical circuit presented in Figure 1.19, elements of which have meaning of the equivalent parameters of a transducer. Each contour corresponds to one of generalized velocities and is coupled to the electrical side of the transducer by ideal electromechanical transformer. The contours are independent as far as the supporting functions θ_i are orthogonal. Otherwise, the equations will be coupled, and the mutual impedances will appear in the contours, between which the coupling exists. The radiation impedances Z_{aci} or/and mechanical impedances $Z_{m\,i}^E$ will be represented in this case as

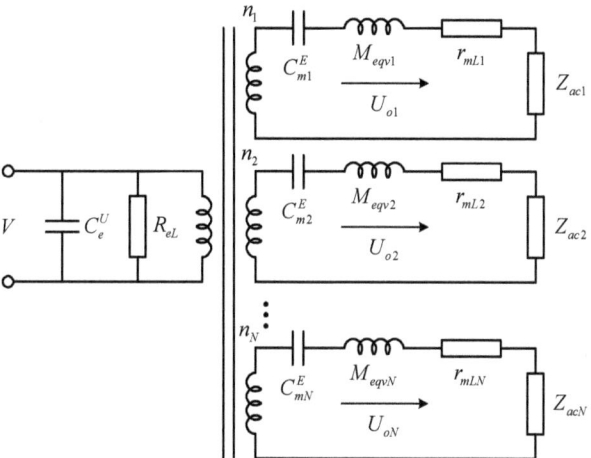

Figure 1.19: Equivalent multi contour electromechanical circuit of a transducer having multiple degrees of freedom in the transmit mode.

$$Z_{aci} = Z_{acii} + \sum_{l \neq i} z_{acil} (U_l / U_i), \quad (1.104)$$

$$Z_{m\,i}^E = Z_{m\,ii}^E + \sum_{l \neq i} z_{m\,il} (U_l / U_i), \quad (1.105)$$

where $Z_{m\,ii}^E$ and Z_{acii} are the self-impedances related to i^{th} mode of vibration, and $z_{m\,il}$, $z_{ac\,il}$ are the mutual impedances between different modes of vibration. Examples of calculating these impedances for particular transducer types are considered in Chapter 4 and Chapter 6. Computational problems of calculating the transducers parameters, and clarity of interpretation of the results obtained depend significantly on how the supporting functions are determined. The right choice of the set of supporting functions is one of the main problems in application of the energy approach to calculating the transducers.

Modification of the equivalent multi contour circuit for the receive mode of operation can be obtained following the procedure used for deriving the circuit in Figure 1.18. As the result the equivalent forces F_{eqvi} that correspond to the modes of vibration θ_i must be inserted in the contours in series with the radiation impedances Z_{aci}.

In conclusion it must be reminded that in this chapter the energy approach is described for calculating parameters of a transducer as an electromechanical device, for which mechanical or acoustic load is known. In order to complete theoretical analysis of the transducer as electroacoustic device, the radiation problem for the transducer must be solved (the Acoustic Subsystem of the transducer must be considered). Analysis of the Acoustic Subsystem of the transducer must result in determining the acoustic load and equivalent force acting on the transducer surface and in determining the spatial distribution of the radiated energy, i.e., in determining the directional properties of the transducer. The general analysis of the Acoustic Subsystems of the transducers is performed in Chapter 6. Both aspects of the transducer treatment in this respect will be illustrated in the next chapter with examples of one degree of freedom transducers.

1.7 References

1. H. Kolsky, *Stress Waves in Solids* (Dover, New York, 1963).
2. D. A. Berlincourt, D. R. Curran, and H. Jaffe, "Piezoelectric and Piezomagnetic Materials and their Function in Transducers," in *Physical Acoustics*, Vol. I, Part A, edited by W. P. Mason (Academic, New York, 1964).
3. N. A. Umov, "Ein Theorem uber die Wechselwirkunglen in Endlichen Entfernungen," Zeitschrift fur Matehematik und Physic, XIX, 97, 1874; Doctoral dissertation "Equations of energy motion in bodies," 1874 (in Russian). It is noteworthy that analogous to Umov's energy flow vector concept was independently developed by John Henry Pointing for electromagnetic fields. In 1884 he introduced the density vector of electromagnetic energy, $S = \left[\bar{E} \times \bar{H} \right]$, which is called Pointing vector (or Umov-Pointing vector).
4. P. M. Morse and H. Feshbach, *Methods of Theoretical Physics*, Part I (McGraw-Hill, New York, 1953).

CHAPTER 2

DESIGNING TRANSDUCERS

2.1 One Degree of Freedom Transducers

Representation of a transducer as a system with one mechanical degree of freedom, i. e., vibrating with fixed velocity distribution in the operating frequency range, proves to be an adequate approximation in many, if not to say in the most, practical cases. Thus, in the case of a transducer operating in vicinity of its resonance frequency (usually for projectors and electromechanical resonators) the mode of vibration of mechanical system can be considered as close to that, which takes place at resonance frequency. Piezoceramic spheres and short rings under uniform electrical excitation and uniform loading can be considered as uniformly vibrating (pulsating) by the symmetry considerations. For the transducers (mostly receivers) that employ flexural vibrations of various mechanical systems approximation for the modes of vibration at frequencies below the first resonance can be obtained in the form of static deflection under the action of correspondingly distributed forces according to Rayleigh's method[1].

In this chapter several examples of widely used transducers having one degree of freedom will be considered with the twofold goal: to obtain the data that are necessary for calculating transducers of this kind, and to formulate the problems of designing the general transducer types with their examples.

2.2 Spherical Transducer

Consider transducer in the shape of a thin spherical shell made of piezoelectric ceramics fully electroded on the inner and outer surfaces and poled in the radial direction (in direction of axis 3 of the crystallographic coordinate system), as shown in Figure 2.1. Under the assumption that $t \ll 2a$, where a is the average radius of the shell, and that the inner and outer surfaces of the shell are free of normal stresses, it can be concluded that in the body of the piezoelement $T_3 = 0$ and $E_3 = V/t$. Due to symmetry $T_1 = T_2$, all the shear stresses are zero, $T_4 = T_5 = T_6 = 0$ and the "working" deformations are the strain in the circumferential direction, $S_1 = S_2$, which will be found as

2.2. Spherical Transducer

$$S_1 = S_2 = \frac{2\pi(a+\xi_o)-2\pi a}{2\pi a} = \frac{\xi_o}{a}. \tag{2.1}$$

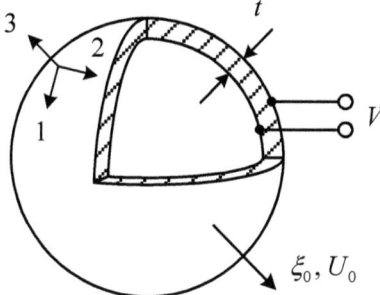

Figure 2.1: Spherical transducer configuration.

The piezoelectric equations with T and E_3 as independent variables in this case become

$$S_1 = S_2 = \left(s_{11}^E + s_{12}^E\right)T_1 + d_{31}E_3, \tag{2.2}$$

$$S_3 = (s_{12}^E + s_{13}^E)T_1 + d_{33}E_3, \tag{2.3}$$

$$D_3 = 2d_{31}T_1 + \varepsilon_{33}^T E_3. \tag{2.4}$$

From these equations we obtain

$$D_3 = \frac{2d_{31}}{s_{11}^E + s_{12}^E} S_1 + \varepsilon_{33}^T \left[1 - \frac{2d_{31}^2}{\varepsilon_{33}^T(s_{11}^E + s_{12}^E)}\right] = D_3^E(S_1) + \varepsilon_{33}^{S_{1,2}} E_3, \tag{2.5}$$

where

$$D_3^E(S_1) = \frac{2d_{31}}{s_{11}^E + s_{12}^E} S_1 \tag{2.6}$$

is the charge density at $E_3 = 0$,

$$\frac{2d_{31}^2}{\varepsilon_{33}^T(s_{11}^E + s_{12}^E)} = k_p^2 \tag{2.7}$$

is the planar coupling coefficient (square) for a piezoceramic material, and

$$\varepsilon_{33}^{S_{1,2}} = \varepsilon_{33}^T(1-k_p^2) \tag{2.8}$$

is the dielectric constant of a piezoelement, "blocked" in the direction of deformations S_1 and S_2. The capacitance of the spherical shell blocked in directions 1 and 2 is

$$C_e^{S_{1,2}} = 4\pi a^2 \varepsilon_{33}^T(1-k_p^2)/t. \tag{2.9}$$

The energy status of the pulsating spherical shell is as follows. The kinetic energy is

$$W_{km} = \frac{1}{2}\int_{\tilde{V}} \rho \dot{\xi}_0^2 d\tilde{V} = \frac{1}{2}\dot{\xi}_0^2 M. \qquad (2.10)$$

Thus, the equivalent mass of the shell is equal to total mass $M_{eqv} = M = 4\pi a^2 t \rho$. The potential energy of the shell at $E_3 = 0$ is

$$W_{pot}^E = \frac{1}{2}\int_{\tilde{V}} (S_1 T_1 + S_2 T_2) d\tilde{V} = \frac{1}{2} 8\pi a^2 t S_1 T_1. \qquad (2.11)$$

After substituting T_1 from Eq. (2.2) at $E_3 = 0$ and $S_1 = \xi_o / a$ we arrive at

$$W_{pot}^E = \frac{1}{2}\xi_0^2 \frac{8\pi t}{s_{11}^E + s_{12}^E}, \qquad (2.12)$$

and the equivalent rigidity of the spherical shell is

$$K_{eqv}^E = \frac{1}{C_{eqw}^E} = \frac{8\pi t}{s_{11}^E + s_{12}^E}. \qquad (2.13)$$

The resonance frequency of the transducer now will be obtained as

$$f_r = \frac{1}{2\pi\sqrt{M_{eqv} C_{eqv}^E}} = \frac{1}{2\pi a\sqrt{\rho s_{11}^E}}\sqrt{\frac{2}{1-\sigma_1^E}}, \qquad (2.14)$$

where it is denoted $\sigma_1^E = -s_{12}^E / s_{11}^E$, as analog of the Poisson's ratio for the piezoelectric ceramics.

For determining the electromechanical energy of the pulsating sphere at first the density of energy must be found in the way analogous to obtaining formula (1.48) as

$$w_{em} = \frac{1}{2}\frac{2d_{31}}{s_{11}^E + s_{12}^E} S_1 E_3 = \frac{1}{2} D_3^E E_3 \qquad (2.15)$$

(the charge density at $E_3 = 0$, D_3^E, is substituted according to relation (2.6)). By integrating over the volume of sphere the total energy will be obtained as

$$W_{em} = \frac{1}{2}\int_{\tilde{V}} D_3^E E_3 d\tilde{V} = \frac{1}{2} n \xi_o V \qquad (2.16)$$

and after substituting D_3^E, $S_1 = \xi_o / a$ and $E_3 = V/t$, we arrive at expression for the electromechanical transformation coefficient

2.2. Spherical Transducer

$$n = \frac{8\pi a d_{31}}{(s_{11}^E + s_{12}^E)}. \tag{2.17}$$

The resistances of the electrical and mechanical losses are according to formulas (1.37) and (1.39)

$$r_{mL} = \frac{1}{\omega C_m^E Q_m}, \quad R_{eL} = \frac{1}{\omega C_e^{S_{1,2}} \tan \delta_e}. \tag{2.18}$$

Now it remains to determine the sound field related parameters of the spherical transducer in order to complete the equivalent circuit of the transducer as an electromechanical device loaded by the radiation impedance Z_{ac}. To complete analysis of the transducer as an electroacoustic device a link must be provided between the transducer surface velocity, $U_0 = \dot{\xi}_0$, and the generated acoustic field, $P_s(r)$, in the transmit mode; and equivalent force F_{eqv} must be determined that characterizes acoustic field as a source of energy for the transducer in receive mode of operation.

The spherical transducer presents an ideal example for illustrating statement of the radiation problems for transducers, because solution for the acoustic field generated by a pulsating sphere is readily available in literature (see, for example, Refs. 2, 3). The general solution for sound pressure radiated by the pulsating sphere is

$$P(r) = \frac{B}{r} e^{-jkr}, \tag{2.19}$$

where quantity B must be determined from the condition that velocity on the surface of the sphere is $U(a) = U_o$, i. e.,

$$-j\omega\rho \left.\frac{\partial P}{\partial r}\right|_{r=a} = U_o. \tag{2.20}$$

After determining the quantity B it will be obtained from Eq. (2.19) that

$$P(r) = \frac{(\rho c)_w}{r} U_0 e^{-j(kr-\pi/2)} \frac{ka^2}{1+jka} e^{jka}. \tag{2.21}$$

Comparing expressions (2.21) and (1.55) we conclude that in the case of the pulsating sphere

$$\chi(\omega) = \frac{ka^2}{1+jka} e^{jka}. \qquad (2.22)$$

After integrating over surface of the sphere in the expression (1.56) at $r_\Sigma = a$ and $\theta(r_\Sigma) = 1$ the radiation impedance will be obtained as

$$Z_{ac} = 4\pi a^2 (\rho c)_w \left[\frac{(ka)^2}{1+(ka)^2} + j\frac{ka}{1+(ka)^2} \right] =$$
$$= r_{ac} + jx_{ac} = (\rho c)_w S_\Sigma (\alpha_r + j\beta_r), \qquad (2.23)$$

where S_Σ is the surface area of the sphere and α_r and β_r are the dimensionless coefficients, dependences of which on the wave dimension of a transducer, $ka = 2\pi a/\lambda$, are shown in Figure 2.2. At values of $ka < 0.3$

$$x_{ac} \approx \omega \rho 4\pi a^3 = \omega M_{ac}, \qquad (2.24)$$

$$r_{ac} \approx \pi(\rho c)_w \frac{(4\pi a^2)^2}{\lambda^2} = \pi(\rho c)_w S_\Sigma^2 / \lambda^2 .. \qquad (2.25)$$

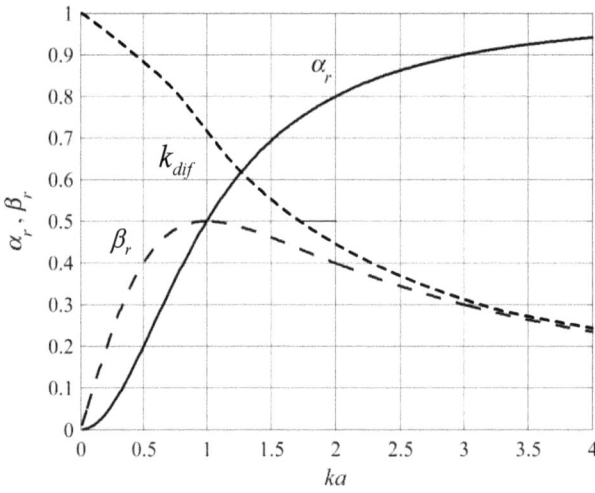

Figure 2.2: The dimensionless coefficients of the radiation impedance of a spherical transducer and diffraction coefficient

In the case that the wave size of the sphere is small ($ka \to 0$), it follows from expression (2.21) that sound pressure generated by the small sphere, P_0, is

$$P_0 = \frac{(\rho c)_w}{r} e^{-j(kr-\pi/2)} U_o ka^2, \qquad (2.26)$$

2.2. Spherical Transducer

or, if to denote $U_o \cdot 4\pi a^2 = U_{\tilde{v}}$ as the volume velocity (or the source strength),

$$P_0 = \frac{(\rho c)_w}{2\lambda r} U_{\tilde{v}} e^{-j(kr-\pi/2)}, \qquad (2.27)$$

where $\lambda = 2\pi/k$ is the wavelength. In this form the formula for sound pressure is valid for all the sources with dimensions much smaller than wavelength (simple sources).

We define the ratio of the sound pressure generated by an arbitrary transducer to the sound pressure generated by the small pulsating sphere having the same volume velocity, as the diffraction coefficient of the transducer in the transmit mode, $k_{dif\,t}$. Using expressions (2.27) and (1.55), we obtain

$$\frac{P(r,\omega)}{P_0(U_{\tilde{v}_r})} = \frac{2\lambda \chi(r,\omega)}{S_\Sigma} = k_{dif\,t}(r,\omega), \qquad (2.28)$$

where S_Σ is the radiating surface of the transducer. In the case of the spherical transducer

$$k_{dif\,t} = \frac{1}{1+jka} e^{jka} = \frac{1}{\sqrt{1+(ka)^2}} e^{j(ka-\arctan ka)}. \qquad (2.29)$$

The function $|k_{dif\,t}|$ is depicted in Figure 2.2. At $ka \gg 1$, $|k_{dif\,t}| \to 1/ka$.

Thus, all the parameters of the equivalent electromechanical circuit of the transducer are determined, and after calculating velocity of vibration of the transducer surface the sound pressure in the acoustic field generated by these vibrations is determined as well. Calculation of the transducer in the transmit mode is completed.

For calculating transducer in the receive mode the value of F_{eqv}, the "mechanomotive force" of the acoustic field as the source of energy for the transducer, must be determined by solving the problem of diffraction of plane acoustic wave by the transducer surface. So far as the radiation problem is solved, this can be done by using the reciprocity principle. We will perform the treatment for transducer surface having a general configuration, because this will be useful for further analysis.

From the general definition (1.61) for the equivalent force

$$F_{eqv} = \int_\Sigma P_u(r_\Sigma) \theta(r_\Sigma) d\Sigma \qquad (2.30)$$

follows that in the case that dimensions of a transducer are small compared with the wavelength of sound, i.e. $P_u(r_\Sigma) \approx P_0$ (P_0 is the sound pressure in the propagating plane wave),

$$F_{eqv} = P_0 \int_\Sigma \theta(r_\Sigma) d\Sigma = P_0 S_{av}. \qquad (2.31)$$

Here the average transducer surface S_{av} is introduced as

$$S_{av} = \int_\Sigma \theta(r_\Sigma) d\Sigma. \qquad (2.32)$$

If the dimensions of the transducer are comparable with the wavelength, the equivalent force may be represented as

$$F_{eqv} = P_0 k_{dif\,r} S_\Sigma, \qquad (2.33)$$

where $k_{dif\,r}$ is the diffraction coefficient in the receive mode, and S_Σ is the radiating surface area of the transducer. In the case of sphere $S_{av} = S_\Sigma = 4\pi a^2$.

Consider the mechanoacoustical system presented in Figure 1.7 that consists of two transducers: the transducer under consideration (#1) with surface Σ, on which the distribution of velocity is specified as $U(r_\Sigma) = U_o \theta(r_\Sigma)$, and the pulsating sphere of small radius a (#2) located at a large distance R from the transducer. One of the formulations of the reciprocity principle is as follows: the pressure at a point 1 (sound pressure $P_{u1}(r_\Sigma)$ acting at the element $d\Sigma$ of the blocked surface Σ) due to a source in point 2 (sphere vibrating with the volume velocity $U_{\dot{v}2} = U_{sph} \cdot 4\pi a^2$) is equal to the pressure at point 2 (at the blocked surface of the sphere) due to a source in point 1 (element $d\Sigma$ of transducer 1 with the volume velocity $U_{\dot{v}1}(r_\Sigma) = U_o \theta(r_\Sigma) d\Sigma$) everything else being equal, i. e.,

$$\frac{P_{u1}(r_\Sigma)}{U_{\dot{v}2}} = \frac{P_{u2}}{U_{\dot{v}1}(r_\Sigma)}. \qquad (2.34)$$

The sound pressure P_0 generated by a pulsating sphere of a small radius in the free field is determined by the expression (2.27). The sound pressure generated in the free field by an elementary source $d\Sigma$ located on a blocked surface Σ, was denoted in formula (1.55) as $P_{d\Sigma}(R, r_\Sigma) = U_o P_\delta(R, r_\Sigma) \theta(r_\Sigma) d\Sigma$, where P_δ is the sound pressure generated by a point source on the transducer surface having unit velocity. As the blocked sphere of a small radius does not disturb the acoustic field, it should be $P_{u2} = P_{d\Sigma}$. After substituting the values $P_{u2} = U_o P_\delta(R, \mathbf{r}_\Sigma) \theta(\mathbf{r}_\Sigma) d\Sigma$, $U_{\tilde{v}_1} = U_o \theta(\mathbf{r}_\Sigma) d\Sigma$ and $U_{\tilde{v}_2}$ determined from formula (2.27) into expression (2.34) written as $P_{u1} U_{\tilde{v}_1} = P_{u2} U_{\tilde{v}_2}$, we obtain

$$P_{u1}(\mathbf{r}_\Sigma) U_o \theta(\mathbf{r}_\Sigma) d\Sigma = \frac{2\lambda R}{\rho c} P_0 e^{j(kR-\pi/2)} U_o P_\delta(R, \mathbf{r}_\Sigma) \theta(\mathbf{r}_\Sigma) d\Sigma. \qquad (2.35)$$

After integrating both parts of this expression over the surface Σ and taking into account formula (1.61) for F_{eqv} and expression (1.55) for the sound pressure generated by the vibrating surface Σ at distance R

$$P(R) = U_o \int_\Sigma P_\delta(R, \mathbf{r}_\Sigma) \theta(\mathbf{r}_\Sigma) d\Sigma = \frac{(\rho c)_w}{R} U_o e^{-j(kR-\pi/2)} \chi(\mathbf{r}, \omega), \qquad (2.36)$$

where $|\mathbf{r}| = R$, we arrive at the following relation

$$F_{eqv} = 2\lambda \cdot \chi(\mathbf{r}, \omega) P_0. \qquad (2.37)$$

Here, P_0 is the sound pressure in the plane wave generated by the pulsating sphere. Given that the pulsating sphere is located at arbitrarily large distance R, P_0 can be considered as the sound pressure of the plane acoustic wave in the free field at the transducer location. The function $\chi(\mathbf{r}, \omega)$ is the known diffraction function obtained from the solution of the radiation problem. By comparing expressions (2.37) and (2.33) the diffraction coefficient for the transducer in the receive mode will be obtained in the form

$$k_{dif\,r} = \frac{2\lambda \chi(\mathbf{r}, \omega)}{S_\Sigma} \qquad (2.38)$$

that coincides with expression (2.28) for the difraction coefficient, $k_{dif\,t}$, introduced for the transducer in the transmit mode. Therefore the distunguishing subscripts t and r will be further omitted. The result obtained is valid for an arbitrary configuration and mode of vibration of the transducer surface. This is manifestation of the principle of resiprocity. The value of the diffraction coefficient for the pulsating sphere is given by expression (2.29).

2.3 Cylindrical Transducers

The cylindrical transducer shown in Figure 2.3 (a) is supposed to be much longer than the wavelength in the operating frequency range. It can be made of piezoelectric ceramic rings in two variants of design: solid radially polarized ring (Figure 2.3 (b)) and segmented ring cemented out of staves having electrodes on their sides and polarized in the circumferential direction (Figure 2.3 (c)).

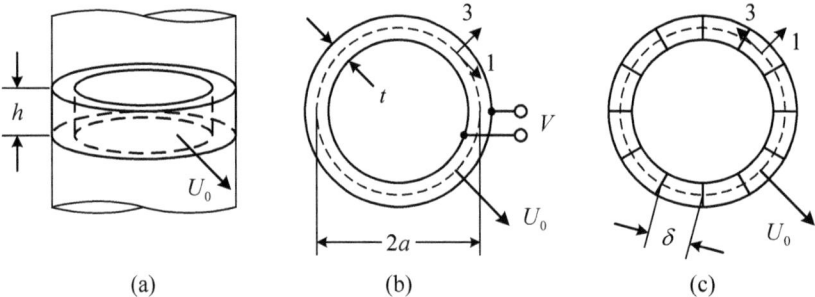

Figure 2.3: (a) Cylindrical transducer made of piezoelectric ceramics rings: h is the height, t is the thickness of the ring, $2a$ is the average diameter, V is the applied voltage, E is the electrical field strength, and U_o is the velocity of reference point; (b) radially poled and (c) circumferentially poled rings.

The following assumptions will be made: the rings are thin and short, i. e., $t, h \ll 2a$; the surfaces of the ring are free of stress, $T_3 = T_2 = 0$, and because of small t and h these stress can be considered zero throughout the volume of the ring; the electric field in the rings is uniform (variation of the electrical field in the radial direction can be neglected because of the small thickness of the rings); the acoustic load is uniform as the radiation is supposed to be axially symmetric. The strength of the electric field in the rings at the radial and circumferential polarizations is $E_3 = V/t$ and $E_3 = V/\delta$, respectively. Under the assumptions made the ring undergoes uniform radial vibrations, i. e., $\xi = \xi_o$. The strain in the circumferential direction will be found as

$$S = \frac{2\pi(a+\xi_o) - 2\pi a}{2\pi a} = \frac{\xi_o}{a} \tag{2.39}$$

and the strains in other directions are idle (as their stress counterparts are zeros) and can be neglected. As the only non-zero stress in the ring is the stress in circumferential direction, the piezoelectric equations in this case can be simplified to the form

$$S_i = s_{ii}^E T_i + d_{3i} E_3, \tag{2.40}$$

$$D_3 = d_{3i} T_i + \varepsilon_{33}^T E_3, \tag{2.41}$$

2.3. Cylindrical Transducers

where $i = 1$ for the radially poled ring, and $i = 3$ for the circumferentially poled. The stress T_i being determined from equation (2.40) is

$$T_i = \frac{1}{s_{ii}^E} S_i - \frac{d_{3i}}{s_{ii}^E} E_3 . \tag{2.42}$$

After substituting this expression into equation (2.41) we obtain

$$D_3 = \frac{d_{3i}}{s_{ii}^E} S_i + \varepsilon_{33}^{S_i} E_3 , \tag{2.43}$$

where $\varepsilon_{33}^{S_i} = \varepsilon_{33}^T(1 - k_{3i}^2)$ is the dielectric constant of a piezoelectric element blocked in the direction of the only "working" deformation S_i ($S_i T_i \neq 0$) and $k_{3i}^2 = d_{3i}^2 / s_{ii}^E \varepsilon_{33}^T$ is the coupling coefficient of piezoelectric ceramics.

In determining the energy status of the ring, the kinetic energy is

$$W_{kin} = \frac{1}{2} \rho \tilde{V} \dot{\xi}_0^2 = \frac{\dot{\xi}_0^2}{2} M , \tag{2.44}$$

where $M = 2\pi ath \cdot \rho$ is the total mass of the ring. The strain (potential) energy of the ring at $E_3 = 0$ is

$$W_{pot}^E = \frac{1}{2} \int_{\tilde{V}} S_i T_i^E d\tilde{V} = \frac{\xi_0^2}{2} \frac{2\pi th}{a s_{ii}^E} , \tag{2.45}$$

and the equivalent compliance of the ring is

$$C_{eqv}^E = \frac{a s_{ii}^E}{2\pi th} . \tag{2.46}$$

Thus, the resonance frequency of a ring is

$$f_r = \frac{1}{2\pi \sqrt{M_{eqv} C_{eqv}^E}} = \frac{1}{2\pi a \sqrt{\rho s_{ii}^E}} . \tag{2.47}$$

According to expression (2.16) the electromechanical energy, W_{em}, can be represented as

$$W_{em} = \frac{1}{2} \int_{\tilde{V}} D_3^E E_3 d\tilde{V} = \frac{1}{2} \xi_0 V n . \tag{2.48}$$

After substituting $D_3^E = d_{3i} S_i / s_{ii}^E$ from (2.43), and values $E_3 = V/t$ and $E_3 = V/\delta$ for variants of the radial and circumferential polarizations we obtain the electromechanical transformation coefficients n_1 and n_3 for these variants

$$n_1 = \frac{2\pi d_{31} h}{s_{11}^E}, \quad n_3 = \frac{2\pi d_{33} h}{s_{33}^E}\frac{t}{\delta}. \tag{2.49}$$

Capacitances of the blocked (in terms of the working deformations S_i) rings are

$$C_e^{S_1} = \varepsilon_{33}^T(1-k_{31}^2)\frac{2\pi a}{t} \text{ and } C_e^{S_3} = \varepsilon_{33}^T(1-k_{33}^2)\frac{2\pi a h t}{\delta^2} \tag{2.50}$$

where k_{3i} are the coupling coefficients of piezoceramics.

The resistances of the electrical and mechanical losses being presented by Eq. (2.18), as quantities inherent in the capacitance and the equivalent compliance, are

$$R_{eL} = \frac{1}{\omega C_e^{S_i} \tan \delta_e}, \quad \tan \delta_e = \frac{1}{Q_e}, \tag{2.51}$$

$$r_{mL} = \frac{\tan \delta_m}{\omega C_m^E}, \quad \tan \delta_m = \frac{1}{Q_m}, \tag{2.52}$$

where Q_e and Q_m are the electrical and mechanical quality factors of the piezoelectric material. Thus, all the parameters of the equivalent electromechanical circuit of a ring resonator (piezoelectric element vibrating without an acoustical load) are determined.

2.3.1 Acoustic Field of the Infinitely Long Cylindrical Transducer

A general analysis of radiation of the cylindrical transducers of different kind is made in Section 6.3. If to suppose that the cylindrical transducer is built from a number of identical rings and it is long compared to the wavelength in water, then the acoustic load can be considered to be uniform by length and the radiation impedance per unit height of the transducer can be found from solution for acoustic field radiated by the infinitely long cylinder. In this case the problem is two-dimensional (independent of the z coordinate). The solution for this problem is described in Ref. 2. The resulting expressions involve Bessel functions. All the information regarding properties of the Bessel functions required for this solution is presented in Ref 2.

The sound pressure in the wave radiating by the infinitely long cylinder is

$$P(r) = A[J_0(kr) - j N_0(kr)]. \tag{2.53}$$

After applying the boundary condition

$$-\frac{1}{j\omega\rho}\frac{\partial P}{\partial r}\bigg|_{r=a} = \frac{k}{j\omega\rho}A[J_1(ka) - j N_1(ka)] = U_o \tag{2.54}$$

2.3. Cylindrical Transducers

(The time depending factor $e^{j\omega t}$ is omitted).

and determining expression for constant A the sound pressure will be obtained in the form

$$P(r) = j\rho c U_0 \frac{J_0(kr) - jN_0(kr)}{J_1(ka) - jN_1(ka)}. \tag{2.55}$$

Here, $J_0(x)$, $J_1(x)$ and $N_0(x)$, $N_1(x)$ are the Bessel and Neumann functions of the zero and first order, respectively.

The radiation impedance per unit height can be determined due to symmetry as

$$Z_{ac}(ka) = 2\pi a \frac{P(a)}{U_0} = j\rho c\, 2\pi a \frac{J_0(ka) - jN_0(ka)}{J_1(ka) - jN_1(ka)}, \tag{2.56}$$

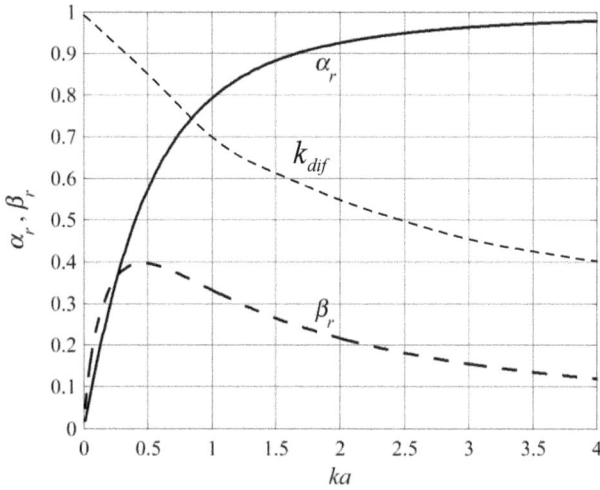

Figure 2.4: Dimensionless coefficients of the components of radiation impedance and diffraction coefficient for the cylindrical transducer.

or

$$Z_{ac}(ka) = r_{ac} + jx_{ac} = \rho c\, 2\pi a [\alpha_r(ka) + j\beta_r(ka)]. \tag{2.57}$$

The non-dimensional coefficients α_r and β_r of radiation impedance are

$$\alpha_r(ka) = \frac{2}{\pi ka} \frac{1}{J_1^2(ka) + N_1^2(ka)}, \quad \beta_r(ka) = \frac{J_0(ka)J_1(ka) + N_0(ka)N_1(ka)}{J_1^{(2)}(ka) + N_1^2(ka)}. \tag{2.58}$$

These functions of ka are plotted in Figure 2.4.

In the "long-wave" approximation (at $ka \ll 1$)

$$\alpha_r(ka) \approx \pi ka/2, \quad \beta_r(ka) \approx ka\ln(1/ka), \qquad (2.59)$$

and per unit height of transducer

$$r_{ac}(ka) \approx \frac{\pi}{2}(\rho c)_w \frac{(2\pi a)^2}{\lambda}, \qquad (2.60)$$

$$x_{ac}(ka) \approx [2\rho_w \pi a^2 \ln(1/ka)]\omega = m_{ac}\omega, \qquad (2.61)$$

where $m_{ac} = 2M_w \ln(1/ka)$ and $M_w = \rho_w \pi a^2$ is the mass of water in the volume of the cylinder per unit length. In the "short-wave" limit (at $ka \to \infty$)

$$P(r) = j\rho c U_0 \sqrt{\frac{2}{\pi kr}} e^{-j(kr-\pi/4)} \frac{1}{J_1(ka) - j N_1(ka)}. \qquad (2.62)$$

Sound pressure generated by the cylinder of very small radius (cylindrical simple source) will be obtained from formula (2.56) taking into consideration that, at $ka \to 0$, $J_1(ka) \to 0$ and $N_1(ka) \to (2/\pi ka)$, as

$$P_0(r)\big|_{ka \ll 1} = \frac{\pi}{2}\rho c\, 2\pi a U_o [J_0(kr) - jN_0(ka)]. \qquad (2.63)$$

Given that $2\pi a U_o = U_{\tilde{V}}$ is the volume velocity per unit length (cylindrical source strength), this formula can be generalized for a long source having small in respect to wavelength cross section (cylindrical simple source) regardless of its geometry and distribution of vibration on the surface as

$$P_0(r)\big|_{ka \ll 1} = \frac{\pi}{2}\rho c U_{\tilde{V}} [J_0(kr) - jN_0(ka)], \qquad (2.64)$$

where $U_{\tilde{V}} = U_o S_{av}$.

Thus, the diffraction coefficient (see Eq. (2.28)) for the cylindrical transducer is

$$k_{dif} = \frac{P(r)_{r \to \infty}}{P_0(r, ka \ll 1)_{r \to \infty}} = j\frac{2}{\pi ka} \frac{1}{J_1(ka) - j N_1(ka)}. \qquad (2.65)$$

And the equivalent force per unit height is (see Eq. (2.33))

$$F_{eqv} = S_\Sigma k_{dif} P_0 = 2\pi a k_{dif} P_0. \qquad (2.66)$$

Here P_0 is the sound pressure in the incident plane wave. The modulus of the diffraction coefficient as function of ka is presented in Figure 2.4. At large ka, $|k_{dif}| \to \sqrt{2/\pi ka}$.

2.3.2 Acoustic Field of the Finite Height Cylindrical Transducer

Real cylindrical transducers have a finite height. Rigorous solution of the radiation problem is available in literature[5] for the case that the cylinder is installed flush with the surface of an infinitely long rigid cylindrical baffle, as shown in Figure 2.5

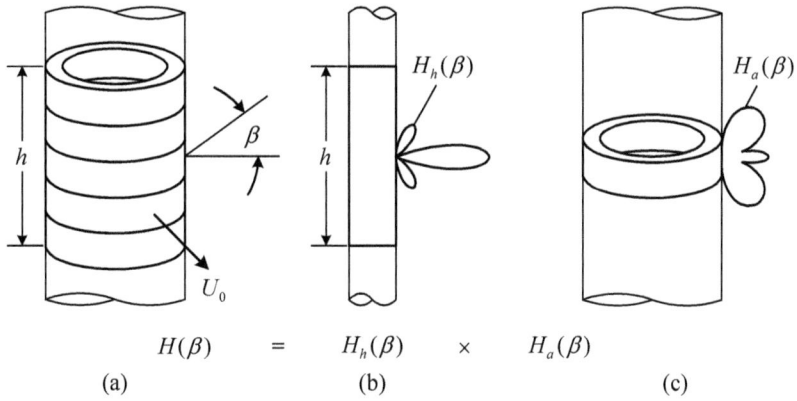

Figure 2.5: Finite-size cylindrical transducer in the infinite rigid baffle and illustration of the product theorem.

It is noteworthy that the results presented below were obtained under the assumption that the surface of the cylinder vibrates uniformly. For the real transducers nonuniformity of radial vibration over the height may exist by two reasons. The radial vibration of the comprising rings over their height may be nonuniform due to coupled vibrations in their mechanical system, if their height to diameter aspect ratio is not small enough (the related issues are considered in Chapter 4). On the other hand, if the wave height of the rings is small and they are connected in parallel the acoustic interaction between the rings may affect distribution of velocities between them significantly especially in the frequency range around their resonance frequency (the related issues are considered in Chapter 6).

At uniform radial vibration by height the sound pressure generated by a cylinder in the far field is

$$P(R,\beta) = -\frac{\rho c h U_o}{\pi \cos\beta H_1^{(2)}(ka\cos\beta)} \cdot \frac{\sin[(kh\sin\beta)/2]}{(kh\sin\beta)/2} \cdot \frac{e^{-jkR}}{R}, \qquad (2.67)$$

where $H_1^{(2)}(x) = J_1(x) - jN_1(x)$ is the Hankel function of the second kind. The sound pressure on the acoustic axis at $\beta = 0$

$$P(R,0) = -\frac{\rho c h U_o}{\pi H_1^{(2)}(ka)} \cdot \frac{e^{-jkR}}{R}. \qquad (2.68)$$

Sound pressure generated by a cylinder of small diameter at $ka \ll 1$ (given that $H_1^{(2)}(ka \ll 1) \approx 2j/\pi ka$) is

$$P(R,\beta)_{ka \ll 1} = \rho c \frac{ka}{2} h U_o \cdot \frac{\sin[(kh\sin\beta)/2]}{(kh\sin\beta)/2} \cdot \frac{e^{-(kR-\pi/2)}}{R}. \qquad (2.69)$$

Sound pressure generated by a short ring at $kh \ll 1$

$$P(R,\beta)_{kl \ll 1} = -\frac{\rho c h U_o}{\pi \cos\beta H_1^{(2)}(ka\cos\beta)} \cdot \frac{e^{-jkR}}{R}, \qquad (2.70)$$

Thus, the directional factor of the transducer, $H(\beta) = P(R,\beta)/P(R,0)$, is

$$H(\beta) = \frac{H_1^{(2)}(ka)}{\cos\beta H_1^{(2)}(ka\cos\beta)} \cdot \frac{\sin[(kh\sin\beta)/2]}{(kh\sin\beta)/2} = H_h(\beta) \cdot H_a(\beta), \qquad (2.71)$$

where

$$H_a(\beta) = \frac{H_1^{(2)}(ka)}{\cos\beta H_1^{(2)}(ka\cos\beta)} \qquad (2.72)$$

is the directional factor of a short ring of radius *a*, and

$$H_h(\beta) = \frac{\sin[(kh\sin\beta)/2]}{(kh\sin\beta)/2} \qquad (2.73)$$

is the directional factor of a thin cylinder of height *h*.

Expression (2.71) is an illustration of the product theorem in the theory of directivity. The above results remain valid for a cylinder without the cylindrical baffle, if its height is greater than the wavelength, $kh \gg 1$ (practically at $h > \lambda$). In the case that the cylinder without the cylindrical baffle has smaller height the radiation problem becomes much more complicated due to the effect of diffraction on its ends. The results obtained for relatively short cylinders are considered in Chapter 6.

2.4 Uniform Bar Transducers

Consider two variants of transducers in the shape of a longitudinally vibrating thin bar: with electrodes on the sides (transverse piezoeffect) and with the electrodes embedded into bar (longitudinal piezoeffect). Their sketches are shown in Figure 2.6. The assumptions are made that the lateral dimensions of the bars are small compared with their length ($w, t \ll l$), and sides and ends of the bars are free of stress. Vibration of the bar is one-dimensional. The only non-zero stress in the bar is the stress in the longitudinal direction, as on the sides of the bar $T_2 = T_3 = 0$ and due to its infinitesimal cross sections they are zeros inside the bar. It is also assumed that in the case of Figure 2.6 (b) the number of segments, N, of which bar is cemented, is large enough for considering that electric field inside the segments practically does not change[6] (it will be shown in Chapter 5 that $N > 6$ is enough). Under this assumption the piezoelectric equations can be used in the form of Eqs. (2.40) and (2.41) that are reproduced here as Eqs. (2.74) and (2.75) with subscripts $i = 1$ for the transverse and $i = 3$ for the longitudinal effects

$$S_i = s_{ii}^E T_i + d_{3i} E_3 , \qquad (2.74)$$

$$D_3 = d_{3i} T_i + \varepsilon_{33}^T E_3 . \qquad (2.75)$$

The solution for free vibration of the thin bar is known.[1] The normal modes of vibration are

$$\xi(x) = \xi_{om} \cos(\pi x m / l), \quad m = 1, 2, \ldots , \qquad (2.76)$$

and strains in the longitudinal direction are

$$S_i = d\xi / dx = -\xi_{om} (\pi m / l) \sin(\pi x m / l) . \qquad (2.77)$$

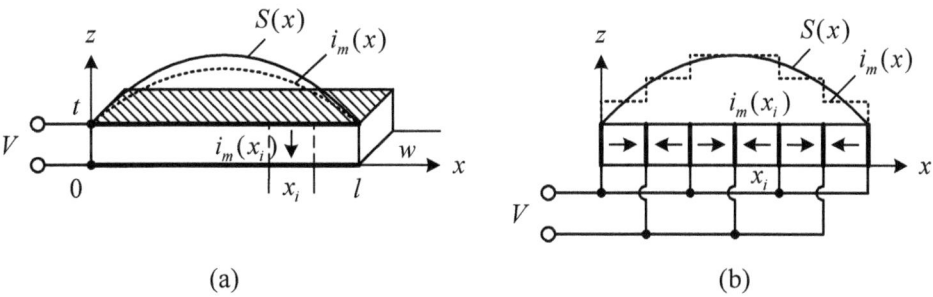

(a)   (b)

Figure 2.6: Longitudinally vibrating piezoceramic bars: (a) side-electroded design, transverse piezoeffect and (b) segmented design, longitudinal piezoeffect.

The sign (–) is in accordance with the convention that tensile strains are positive (they correspond to outside displacements of the ends). In the context of determining the equivalent parameters the sign can be omitted.

In general, the vibrating bar must be treated as a system with multiple degrees of freedom, if to consider its behavior in a broad frequency range. But it can be assumed that in the vicinity of each natural resonance frequency the corresponding resonance mode of vibration dominates, and the bar can be considered as one degree of freedom system. Equivalent parameters of the bars at different normal modes of vibration will be found from expressions for energies associated with the vibrations.

Thus, the equivalent mass of the bar for a given normal mode, being found from the expression for kinetic energy

$$W_{kin} = \frac{1}{2}\int_{V} \rho \dot{\xi}_{om}^{2} d\tilde{V} = \frac{1}{2}\dot{\xi}_{om}^{2} w\rho \int_{0}^{l} \cos^{2}(m\pi x/l)dx = \frac{1}{2}\dot{\xi}_{om}^{2} M_{eqvm}, \quad (2.78)$$

is $M_{eqvm} = M/2$, where $M = wtl\rho$ is the total mass of the bar.

The equivalent rigidity of the resonator will be found from the expression for potential energy, in which the strain is determined by formula (2.77) and the stress at $E_3 = 0$ is $T_i^E = S_i / s_{ii}^E$, as follows from Eq. (2.74). Thus,

$$W_{pot}^E = \frac{1}{2} wt \int_{0}^{l} T_i^E S_i dx = \frac{\xi_{om}^{2}}{2} \frac{\pi^2 wtm^2}{2 s_{ii}^E l} = \frac{\xi_{om}^{2}}{2} K_{eqvm}^E \quad (2.79)$$

and

$$K_{eqvm}^E = 1/C_{eqvm}^E = \frac{\pi^2 wtm^2}{2 s_{ii}^E l}. \quad (2.80)$$

Note that this value of the rigidity is valid for the bar with full size electrodes that cover all the side surfaces (with all segments active in the segmented design). In the case that the electrodes are partial different values for elastic constant must be used, when integrating over parts of the bar free of electrodes. If these parts were not polarized, the Young's modulus of non-polarized ceramics must be used. If they were polarized and the electrodes were removed afterwards, the elastic constant s_{ii}^D (at condition that charge density is zero) should be used. This must be taken

2.4. Uniform Bar Transducers

into consideration in determining the accurate values of the resonance frequencies of the bars by formula

$$f_{rm} = 1/2\pi \sqrt{M_{eqvm} C^E_{eqvm}} = m/2l\sqrt{\rho s^E_{ii}}, \qquad (2.81)$$

where $\left(1/\sqrt{s^E_{ii}\rho}\right) = c^E_i$ is the velocity of propagating the longitudinal vibration in a piezoceramics bar transversely (at $i = 1$) or longitudinally (at $i = 3$) polarized. After the capacitances of the bars is determined, as

$$C^{S_1}_{el} = \varepsilon^T_{33}(1-k^2_{31})wl/t \text{ and } C^{S_3}_{el} = \varepsilon^T_{33}(1-k^2_{33})wtN/\Delta = \varepsilon^T_{33}(1-k^2_{33})wtl/\Delta^2, \qquad (2.82)$$

the resistances of the electrical and mechanical losses may be found by formulas (2.51).

Using the general formula for the electromechanical energy, after substituting D^E_3 and strain S_i by formulas (2.43) and (2.77), respectively, will be obtained

$$W_{em} = \frac{1}{2}\int_{\tilde{V}} D^E_3 E_3 d\tilde{V} = \frac{1}{2}\frac{d_{3i}}{s^E_{ii}}wt\int_0^l S_i E_3(x)dx =$$

$$= \frac{1}{2}\xi_{om} \frac{\pi m d_{3i}}{ls^E_{ii}}wt\int_0^l (-)\sin(\pi mx/l)E_3(x)dx = \qquad (2.83)$$

$$= \frac{1}{2}\xi_{om} V n_m.$$

From this expression follows that in the case of the side-electroded bar (at $E_3(x) = V/t$) with full size unipolar electrodes, the electromechanical transformation coefficient will be

$$n_m = 2d_{31}w/s^E_{11} \text{ at } m = 1,3,... \text{ and } n_m = 0 \text{ at } m = 2,4,.... \qquad (2.84)$$

In the case of segmented bar with all the segments having length δ active and connected as shown in Figure 2.6 (b) (at $E_3(x) = V/\delta = VN/l$)

$$n_m = 2d_{33}wt/\delta s^E_{33} = 2d_{33}wtN/ls^E_{33} \text{ at } m = 1,3,... \text{ and } n_m = 0 \text{ at } m = 2,4,.... \qquad (2.85)$$

Thus, in the case of fully active bars only the odd modes of natural vibrations of the bar can be excited electrically. This fact has a simple physical explanation that is illustrated by plots of the natural mode shapes in Figure 2.7 for the case of the transverse piezoeffect. If to assume that the bar vibrates under an external action in an even mode, then the charge on the electrodes, which is proportional to the strain, will be averaged to zero.

At the odd modes the total charge remains finite even for the high modes, though it drops relative to those for the first mode. Imagine now that the electrodes are divided into two equal parts, and the halves of the electrodes are connected electrically in the opposite phase.

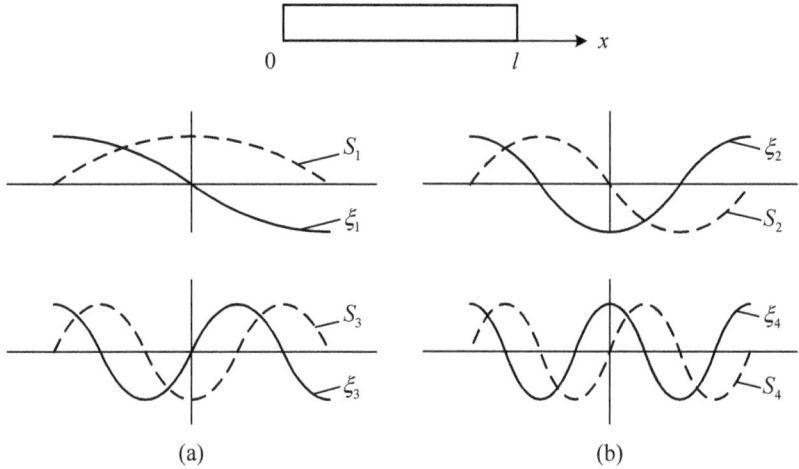

Figure 2.7: The normal modes of longitudinal vibrations of a bar with free ends: (a) odd modes, (b) even modes.

Then $E_3(x) = V/t$ at $0 < x < l/2$ and $E_3(x) = -V/t$ at $l/2 < x < l$. In this case

$$n_m = 4d_{31} w / s_{11}^E \text{ at } m \text{ even and } n_m = 0 \text{ at } m \text{ odd}. \qquad (2.86)$$

The same explanation regarding excitation of modes with reference to Figure 2.7 remains intact. Thus, by switching parts of the electrodes on the bar surface it is possible to excite different modes of natural vibration and to change the resonance frequencies of the bar. The most usable are the first and third modes of vibration. It is evident from Figure 2.7 that transformation coefficient for the third mode can be increased by dividing the electrodes in three parts and connecting the neighboring parts in the opposite phase. In this case the destructive effect of changing sign of the stress along the bar will be eliminated. The corresponding calculation using expression (2.83) results in

$$n_3 = 6 d_{31} w / s_{11}^E. \qquad (2.87)$$

By connecting the halves of the electrodes in antiphase the even modes of vibration will be generated.

It is noteworthy that the equivalent parameters must be used for calculating transducer and for predicting its quality only collectively. Their individual values depend on how the position of reference point is chosen. But some combinations of the equivalent parameters don't depend on a choice of the reference point. They are inherent in the transducer configuration and can be considered as transducer's figures of merit.

2.4.1 Effective Coupling Coefficient of a Transducer

One of important properties of a transducer, which is directly linked to its quality as an energy converter, is the effective coupling coefficient that can be defined as[7]

$$k_{eff}^2 = \frac{\text{mechanical energy stored in the mode of vibration}}{\text{total input energy}} \qquad (2.88)$$

at $\omega \to 0$.

It will be obtained after integrating expressions (1.46) and (1.52) over volume of the bar transducer that the total electrical energy supplied is

$$W_{el} = W_e^S + W_{em} = \frac{1}{2}\left(V^2 C_e^{S_i} + V\xi_{om} n_m\right); \qquad (2.89)$$

the energy stored in mechanical form in the mode under consideration is

$$W_{pot\,m}^E = \frac{1}{2}\xi_{om}^2 K_{eqv\,m}^E = W_{em\,m} = \frac{1}{2}V\xi_{om} n_m, \qquad (2.90)$$

where from

$$V/\xi_{om} = K_{eqv\,m}^E / n_m. \qquad (2.91)$$

After using expressions (2.89)–(2.91) and definition (2.88), we arrive at the formula for k_{eff}^2

$$k_{eff}^2 = \frac{1}{(K_{eqv\,m}^E C_e^{S_i}/n_m^2)+1} \qquad (2.92)$$

that can be represented for brevity as

$$k_{eff}^2 = \frac{\alpha_c}{\alpha_c+1}, \text{ where } \alpha_c = \frac{n_m^2}{K_{eqv\,m}^E C_e^{S_i}} = \frac{n_m^2 C_{eqv\,m}^E}{C_e^{S_i}}. \qquad (2.93)$$

Formulas (2.93) for the effective coupling coefficient and coefficient α_c, which will be used further for denoting the combination of equivalent electromechanical parameters, being obtained regarding the side-electroded transversely poled and for the segmented longitudinally poled bar, are valid for all the one-dimensional piezoelements (for the segmented bar under the assumption that number of segments on the half wave length of deformation is large enough). Detailed treatment of the effective coupling coefficients of piezoceramic bodies under general assumptions regarding their modes of polarization and distribution of deformation will be done in Chapter 5.

The effective coupling coefficients of a bar with full size unipolar electrodes (with all the segments connected as shown in Figure 2.6 (b)) for the first and third modes of vibration being determined by formulas (2.93) are

$$\alpha_{cm} = \frac{k_{3i}^2}{1-k_{3i}^2} \cdot \frac{8}{\pi^2 m^2}, \quad k_{eff\,m}^2 = \frac{1}{1+\pi^2 m^2 (1-k_{3i}^2)/8k_{3i}^2}. \qquad (2.94)$$

Thus, for the bar made from PZT-4 ceramics ($k_{31} = 0.33$ and $k_{33} = 0.7$) $k_{eff\,1} = 0.3$ and 0.66; $k_{eff\,3} = 0.1$ and 0.28. In the case that the electrodes are divided in three parts and the neighboring parts are connected in opposite phase with transformation coefficient determined by formula (2.87) will be obtained that $k_{eff\,3} = 0.3$ and 0.66 (it is easy to make sure that in this case $k_{eff\,1} = 0$). The same result $k_{eff\,2} = 0.3$ and 0.66 will be also obtained for the case that the electrodes are divided into two parts and are connected in opposite phase.

2.5 Mass Loaded Bar Transducer

Consider transducer that has symmetrical configuration shown in Figure 2.8. Mechanical system of the transducer consists of piezoceramic bar loaded by massive identical passive parts.

Due to symmetry the ends of the transducer vibrate with equal velocity $\dot{\xi}_o = U_o$ that fully determines the energy status of the mechanical system and allows considering it as one degree of freedom system. This is a particular case of the transducers of general type, mechanical systems of which are combined of the bars having different lengths and cross section areas. Under the assumptions that $kl_c \ll 1$ and $kL_h \ll 1$ (in reality at the conditions that $\sin kl_c \approx kl_c - (kl_c)^3/6$ and $\sin kl_h \approx kl_h$, as this will be shown in Chapter 10, deformations of the heads can be neglected (they vibrate as a whole with velocity $\dot{\xi}_o$), distribution of displacements in the piezoceramic bar, $\xi(x)$, can be assumed to be linear and the strain and stress to be constant,

2.5. Mass Loaded Bar Transducer

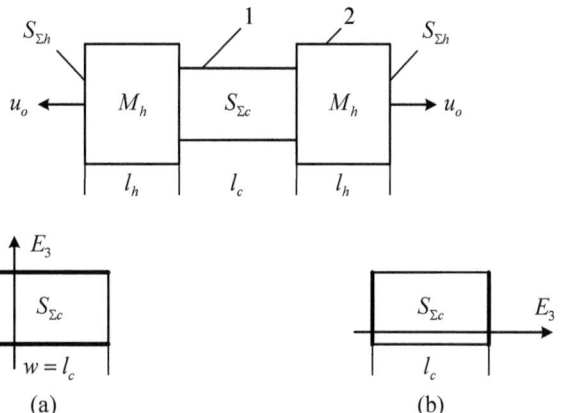

Figure 2.8: Configuration of the symmetrical mass loaded bar transducer: 1 – piezoceramic bar with cross section area $S_{\Sigma c}$, 2 – passive "heads" having cross section area $S_{\Sigma h}$ and mass M_h. In the cases (a) and (b) the transverse and longitudinal piezoeffects are used, respectively.

$$\xi(x) = 2\xi_o x/l_c, \quad S_i = 2\xi_o/l_c, \quad T_i = 2\xi_o/l_c s_{ii}^E . \tag{2.95}$$

The set of piezoelectric equations, Eqs. (2.40) and (2.41), is applicable. Expressions for the energies involved and for the equivalent parameters are as follows. The kinetic energy is

$$W_{kin} = \frac{1}{2}\dot{\xi}_o^2 \left(2M_h + \rho_c S_{\Sigma c} \frac{8}{l_c^2} \int_0^{l_c/2} x^2 dx \right)$$
$$= \frac{1}{2}\dot{\xi}_o^2 (2M_h + m_c/3) = \frac{1}{2}\dot{\xi}_o^2 M_{eqv}, \tag{2.96}$$

where ρ_c is the density of ceramics and m_c is the mass of the bar.

The potential energy is

$$W_{pot} = \frac{1}{2} S_i T_i \cdot S_{\Sigma c} l_c = \frac{1}{2}\xi_o^2 K_{eqv}^E, \quad K_{eqv}^E = 1/C_{eqv}^E = 4S_{\Sigma c}/l_c s_{ii}^E . \tag{2.97}$$

The electromechanical energy is

$$W_{em} = \frac{1}{2} D_3^E E_3 \cdot S_{\Sigma c} l_c = \frac{1}{2}\xi_o Vn . \tag{2.98}$$

Analogous to Eq. (2.43) $D_3^E = d_{3i} S_i / s_{ii}^E$; $E_3 = V/t$ for the case of the transverse effect (a) and $E_3 = V/l_c$ for the case of the longitudinal effect with solid end-electroded bar (b). Thus,

$$n_{(a)} = \frac{2wd_{31}}{s_{11}^E}, \quad n_{(b)} = \frac{2S_{\Sigma c} d_{33}}{l_c s_{33}^E} . \tag{2.99}$$

The clamped capacitances are (with dielectric constant $\varepsilon_{33}^S = \varepsilon_{33}^T(1-k_{3i}^2)$ analogous to Eq.(2.43))

$$C_{e(a)}^{S_1} = \varepsilon_{33}^T(1-k_{31}^2)wl_c/t, \quad C_{e(b)}^{S_3} = \varepsilon_{33}^T(1-k_{33}^2)S_{\Sigma c}/l_c. \tag{2.100}$$

It is easy to check using formulas (2.93) that the effective coupling coefficients are $k_{eff(a)} = k_{31}$ and $k_{eff(b)} = k_{33}$, i. e., the same as the coupling coefficients of the piezoceramics. Actually, this result is obvious, because the piezoelement possesses all the mechanical (potential) energy and its deformation is uniform.

The resonance frequency of the transducer being determined by formula $f_r = 1/2\pi\sqrt{M_{eqv}C_{eqv}^E}$ is

$$f_r = \frac{1}{\pi l_c \sqrt{\rho_c s_{ii}^E}} \sqrt{\frac{m_c}{2M_h + m_c/3}}. \tag{2.101}$$

Usually, $(m_c/3) \ll 2M_h$ and the formula reduces to

$$f_r = \frac{c_c}{\pi l_c} \sqrt{\frac{m_c}{2M_h}}. \tag{2.102}$$

Let L be the length of a uniform bar that has the same resonance frequency, $L = c_c/2f_r$. The ratio of the lengths l_c and L is

$$\frac{l_c}{L} = \frac{2}{\pi}\sqrt{\frac{m_c}{2M_h}}. \tag{2.103}$$

Thus, the same resonance frequency can be obtained with much shorter piezoceramic bar by employing the mass loaded design. This is one of the reasons for using such design. Another reason is that this can be regarded as a way of matching a bar transducer with acoustic load by changing radiation resistance per unit area of the bar cross section. To show this, consider dependence of the radiation impedance of the end surface of a bar from its wave dimensions. We will assume for qualitative estimating the effects of transducer loading that the radiating surface presents a uniformly vibrating circular piston installed flush with the surface of the absolutely rigid plane. This may be considered as an imitation of the situation that the transducer operates in a flat array of a large wave size, as shown in Figure 2.9, and its self-radiation impedance is considered. (In reality an acoustic interaction between transducers in arrays exists and may

2.5. Mass Loaded Bar Transducer

affect the radiation impedance of a single transducer. These issues will be considered in Chapter 6.)

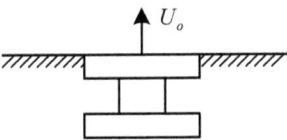

Figure 2.9: Bar transducer operating in a rigid baffle of a large wave size.

The radiation impedance of a circular piston vibrating in the rigid plane is known from literature (see, for example, Ref. 8) as

$$Z_{ac} = r_{ac} + jx_{ac} = (\rho c)_w S_\Sigma (\alpha_r + j\beta_r), \quad (2.104)$$

where the dimensionless coefficients of the components of radiation impedance are

$$\alpha_r(ka) = 1 - \frac{J_1(2ka)}{ka} = \frac{(2ka)^2}{2^2 1!2!} - \frac{(2ka)^4}{2^4 2!3!} + \ldots, \quad (2.105)$$

$$\beta_r(ka) = \frac{S_1(2ka)}{ka} = \frac{4}{\pi}\left[\frac{2ka}{3} - \frac{(2ka)^3}{3^2 \cdot 5} + \frac{(2ka)^5}{3^2 \cdot 5^2 \cdot 7} - \ldots\right], \quad (2.106)$$

Here J_1 and S_1 are the Bessel and Struve functions of the first order. In the case that $ka \ll 1$ (low frequency approximation)

$$\alpha_r \approx (ka)^2 / 2, \quad \beta_r \approx 8(ka)/3\pi, \quad (2.107)$$

At $ka \gg 1$ $\alpha_r \to 1$, $\beta_r \to 0$, and $Z_{ac} \to (\rho c)_w S_\Sigma$. This will be impedance per surface area of transducer operating in a large array vibrating in phase. The dimensionless coefficients of radiation impedance are presented in Figure 2.10.

The diffraction coefficient for the piston-like radiating surface flush with the rigid baffle of a large size is $k_{dif} = 2$, because the sound pressure of the incident wave doubles on the rigid plane. The assumption can be made for rough estimations that maximum dimensions of the radiating surface of a single bar projector in array are $\lambda/2$, i. e., $ka \approx \pi/2$. At this value of ka $\alpha_r \approx 0.81$ and $\beta_r = 0.67$. Thus, the radiation resistance per cross section of uniform bar is $r_{acc} = 0.81(\rho c)_w$. For the mass loaded bar that has radiating surface of the same size the total radiation resistance is $r_{ac\Sigma h} = 0.81(\rho c)_w S_{\Sigma h}$.

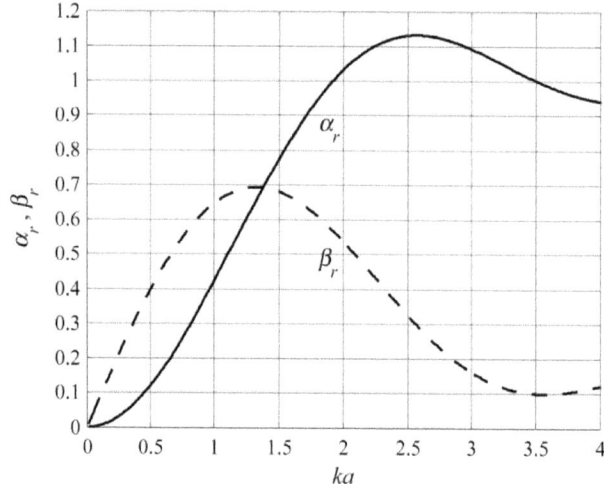

Figure 2.10: Non-dimensional coefficients of radiation impedance of a circular piston vibrating in the infinite rigid plane vs. its wave size.

One of important characteristics of a transducer in operation around the resonance frequency is the quality factor of its mechanical system acoustically loaded, Q_{mw}. By definition

$$Q_{mw} \approx (\omega_r M_{eqv})/r_{ac} = 1/(\omega_r C_{eqv} r_{ac}). \tag{2.108}$$

Compare the quality factors of the uniform and mass loaded bar transducers, $Q_{mw\,u}$ and $Q_{mw\,mL}$ that have equal resonance frequencies and areas of radiating surfaces, i. e., equal radiation resistances. The equivalent compliances of transducers of these types, $C_{eqv\,u}$ and $C_{eqv\,mL}$, are determined by formulas (2.80) and (2.97), respectively. Thus,

$$\frac{Q_{mw\,mL}}{Q_{mw\,u}} = \frac{C_{eqv\,u}}{C_{eqv\,mL}} = \frac{8}{\pi^2} \cdot \frac{L}{l_c} \cdot \frac{S_{\Sigma c}}{S_{\Sigma h}}, \tag{2.109}$$

where from follows that the quality factor of the mass loaded transducer can be transformed (reduced) by changing proportions of its mechanical system. For example, if $(L/l_c) = 2$ and the radius of head is twice the radius of ceramics, $Q_{mw\,mL} = 0.4 Q_{mw\,u}$. Effects of such transformation of the quality factor will be discussed in the next chapter.

2.6 Flexural Type Transducers

Consider transducers with mechanical systems in the shape of rectangular and circular plates, and the cantilever beam that are shown in Figure 2.11.

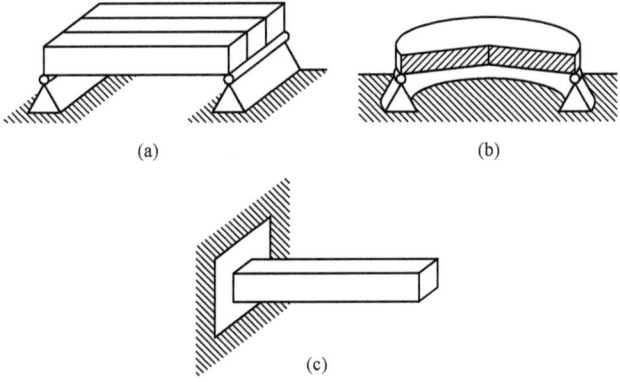

Figure 2.11: Mechanical systems of the flexural type transducers: (a) rectangular plate, (b) circular plate, (c) cantilever beam.

Analysis of transducers of these types in this chapter will be restricted by their applications as receivers (hydrophones, accelerometers) of small size in comparison with the wavelengths, in which capacity they are predominantly used. Therefore, vibrations of the mechanical systems in the frequency range below their first resonance are of interest. The assumption is that the mechanical systems are thin, $t \ll l$, and the rectangular plate and cantilever are comprised of the beams having small width, $w \ll l$. Therefore, it is sufficient to consider vibration of the beams. For the circular plates the assumption is that $t \ll a$ (at least $t/a < 1/5$). Thus, in all the cases elementary theory of bending is applicable. Under these assumptions the modes of vibration of the mechanical systems can be used in the form of static deflection under action of uniformly distributed forces according to Rayleigh's method. The modes of static deflection for the mechanical systems shown in Figure 2.11 can be found in Ref. 9 and in Section 4.5.9. The boundary conditions for the beams and circular plate will be assumed to be simply supported, and the end of the cantilever is clamped.

2.6.1 Rectangular Beam Transducer

The beam is bimorph, i. e., it is cemented of two identical piezoceramic layers poled through their thickness. The way the poling vectors are directed in the layers and the way the electrodes on their surfaces are connected to make the plates active under the flexural deformation are shown in Figure 2.12. Because of the small thickness and width of a beam, it can be assumed that $T_2 = T_3 = 0$, and the only active stress is T_1. Therefore, the appropriate piezoelectric equations are

$$S_1 = s_{11}^E T_1 + d_{31} E_3, \qquad (2.110)$$
$$D_3 = d_{31} T_1 + \varepsilon_{33}^T E_3. \qquad (2.111)$$

After substituting expression for T_1 from Eq. (2.110) into Eq. (2.111) it will be obtained

$$D_3 = \frac{d_{31}}{s_{11}^E} S_1 + \varepsilon_{33}^T (1 - k_{31}^2) E_3. \qquad (2.112)$$

Let displacements of the surface of the bean be $\xi(x) = \xi_o \theta(x)$. The strains S_1 in the thin layer located at distance z from the neutral surface of the plate in Figure 2.12 (b) (neutral is the surface that remains unstretched under flexure) can be determined from the relation

$$S_1 = \frac{(R_1 + z)d\varphi - R_1 d\varphi}{R_1 d\varphi} = \frac{z}{R_1} = -z\xi_o \frac{d^2\theta}{dx^2}, \qquad (2.113)$$

where

$$R_1 = -1/ (d^2\xi/dx^2) \qquad (2.114)$$

is the radius of curvature of the bent beam in zx plane.

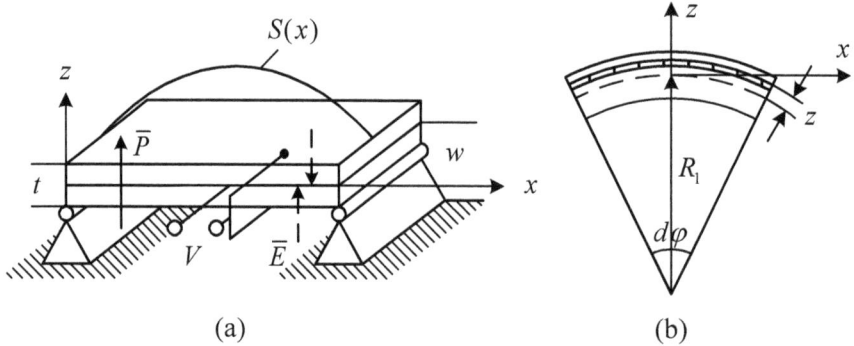

Figure 2.12: Piezoelement represented by a bimorph beam.

Note that the sign in Eq. (2.112) is in accordance with adopted sign convention. The curvature is negative, the layers at $z > 0$ experience tension, which is positive by sign convention, and bending is produced by positive moments. From Eq. (2.110)

2.6. Flexural Type Transducers

$$T_1^E = \frac{1}{s_{11}^E} S_1 = Y_1^E S_1, \qquad (2.115)$$

where notation $Y_1^E = 1/s_{11}^E$ is introduced for the Young's modules of a piezoceramic material.

The expression for the potential energy of the deformed beam is

$$W_{pot}^E = \frac{1}{2}\int_V S_1 T_1^E d\tilde{V} = \frac{\xi_o^2}{2} Y_1^E w \int_0^l \int_{-t/2}^{t/2} z^2 \left(\frac{d^2\theta}{dx^2}\right)^2 dz dx = \frac{\xi_o^2}{2} K_{eqv}^E, \qquad (2.116)$$

where the equivalent rigidity of the plate is

$$K_{eqv}^E = Y_1^E \frac{wt^3}{12}\int_0^l \left(\frac{d^2\theta}{dx^2}\right)^2 dx. \qquad (2.117)$$

Here $wt^3/12 = J_y$ is the moment of inertia of the beam cross section with respect to the y-axis in the neutral plane. The kinetic energy of the beam is

$$W_{kin} = \frac{1}{2}\int_V \rho \dot{\xi}^2 d\tilde{V} = \frac{\dot{\xi}_o^2}{2} wt\rho \int_0^l \theta^2(x)dx = \frac{\dot{\xi}_o^2}{2} M_{eqv}, \qquad (2.118)$$

where

$$M_{eqv} = \rho tw \int_0^l \theta^2(x)dx = \rho t S_{eff} \qquad (2.119)$$

is the equivalent mass and

$$S_{eff} = w \int_0^l \theta^2(x)dx \qquad (2.120)$$

is the effective surface area of the mechanical system.

The electromechanical energy associated with the flexural deformation is

$$W_{em} = \frac{1}{2}\int_V D_3^E E_3(z)d\tilde{V} = \frac{1}{2}\frac{d_{31}}{s_{11}^E} w\xi_o \int_0^l \frac{d^2\theta(x)}{dx^2}dx \int_{-t/2}^{t/2} (-)zE_3(z)dz = \frac{1}{2}\xi_o Vn. \qquad (2.121)$$

Here D_3^E is substituted from Eq. (2.112). From this expression follows that the flexural vibration can be generated electromechanically in the case that the electric field changes sign in halves of the beam. The sign of electric field depends on the relative directions of vectors of electric field E_3 and field of polarization P: it is positive if their directions coincide, and negative otherwise. Two variants of the piezoelement design for generating the bending moment are possible that correspond to the parallel and series connection of its halves. They are shown in Figure 2.13. In the variant of Figure 2.13 (a) that is adopted also in Figure 2.12 (parallel connection) $E_3 = 2V/t$, in the variant (b) (series connection) $E_3 = V/t$. The halves of the beams further are assumed to be connected in parallel.

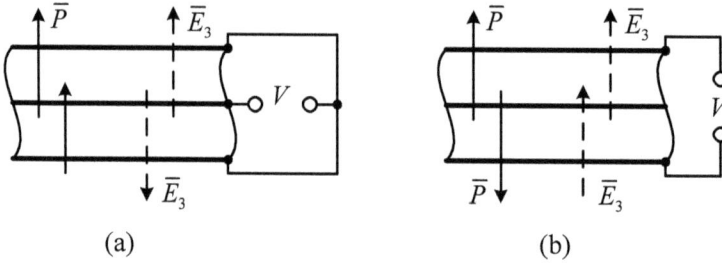

Figure 2.13: Variants of electrical connection of halves of the bimorph beam.

The expression for the electromechanical transformation coefficient that follows from Eq. (2.121) for the parallel connection of halves of the beams is

$$n = -\frac{wtd_{31}}{2s_{11}^E}\left(\frac{d\theta}{dx}\bigg|_{x=l} - \frac{d\theta}{dx}\bigg|_{x=0}\right). \tag{2.122}$$

Thus, the value of transformation coefficient depends on the slopes of the mode shape of displacement on the ends of the electrodes. The sign (–) is related to sign convention and in the context of calculating parameters of a single transducer it can be omitted.

For calculating capacitance of the beam the value of dielectric permittivity $\varepsilon_{ee}^{S_1} = \varepsilon_{33}^T(1-k_{31}^2)$ of the piezoelement clamped in the x direction must be used according Eq. (2.112), thus,

$$C_e^{S_1} = \varepsilon_{33}^T(1-k_{31}^2)\frac{4wl}{t}. \tag{2.123}$$

2.6.2 Cantilever Beam Transducer

All the general expressions for the transducer equivalent parameters are the same as for the simply supported beam, but the reference point has to be chosen at $x = l$, at which point $\theta(l) = 1$. Thus, the reference velocity will be $U_l = \dot{\xi}_l$. The main difference in calculating the parameters is due to different expressions for the mode shapes of vibration.

2.6. Flexural Type Transducers

If the beam is used as mechanical system of accelerometer, the support vibrates with acceleration $\ddot{\xi}_s$, as shown in Figure 2.14. The equivalent force that generates vibration of the beam may be determined in this case as follows. In the range of frequencies below the resonance, in which the accelerometers are usually used, the mode of vibration of a beam can be represented as superposition of its movement together with the support and vibration of the beam relative to the support, namely, as

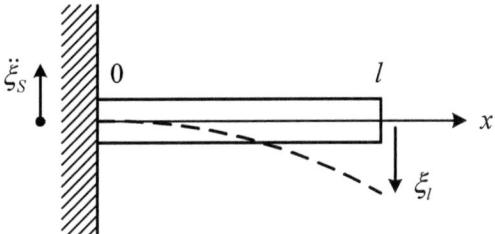

Figure 2.14: Configuration of the cantilever beam transducer.

$$\xi(x) = \xi_s - \xi_o \theta(x) \approx \xi_s, \qquad (2.124)$$

It is taken into consideration in this relation that below the resonance $\xi_o \theta(x) \ll \xi_s$. Thus, the mechanical system vibrates under action of the distributed inertia force with density $f_{in} = \rho t \ddot{\xi}_s$. The equivalent force that generates deformation of the mechanical system is according to formula (1.17)

$$F_{eqv} = w \int_0^l f_{in} \theta(x) dx = \rho wt \ddot{\xi}_s \int_0^l \theta(x) dx = \rho t S_{av} \ddot{\xi}_s. \qquad (2.125)$$

Here,

$$S_{av} = w \int_0^l \theta(x) dx \qquad (2.126)$$

is the average surface area of the mechanical system, one of the equivalent parameters of a transducer. Note that in deriving the equivalent force by formula (2.125) the acceleration was assumed to act in the normal to surface of the beam direction. In general acceleration is vector $\ddot{\boldsymbol{\xi}}_s$ that can be directed at some angle φ to the surface. If $\boldsymbol{n}$ is the unit vector normal to the surface, then in formula (2.125) it must be

$$\ddot{\xi}_s = \ddot{\boldsymbol{\xi}}_s \cdot \boldsymbol{n} = |\ddot{\boldsymbol{\xi}}_s| \cos \varphi \quad \xi_s = \boldsymbol{\xi} \cdot \boldsymbol{n} = |\boldsymbol{\xi}| \cos \varphi. \qquad (2.127)$$

This means that the accelerometer possesses a directional factor that is characterized by the figure of eight pattern.

To complete the calculation of the equivalent parameters, the mode of displacement $\theta(x)$ has to be determined. Generally, the mode of vibration has to be found by solving differential equation of motion of a mechanical system under particular boundary conditions. In the case that exact expressions for the mode shapes are not available, extension of the Rayleigh's method can be applied. According to this method the mode of static deflection of a mechanical system under the action of a distributed load can be used to determine its first resonance frequency. The same mode of displacement can be used for calculating the equivalent parameters of a transducer employing the mechanical system at resonance and in the frequency range below the first resonance. The modes of static displacements under uniform loading are available for different mechanical systems from works on the strength of materials, for example, from Ref. 9. In particular, for a beam with simply supported ends (with displacements and bending moments on the ends zeros)

$$\theta(x) = (16/5!)\left(x - 2x^3/l^2 + x^4/l^3\right), \tag{2.128}$$

and for a cantilever beam with one clamped end (displacement and slope of the displacement curve are zeros at this end)

$$\theta(x) = 2(x/l)^2 (1 - 2x/3l + x^2/6l^2). \tag{2.129}$$

Expressions for the equivalent parameters obtained from the general formulas after using the modes of vibration (2.128) and (2.129) for the simply supported beam and for the cantilever are as follows. For the simply supported beam

$$K_{eqv}^E = 4\frac{wt^3}{l^3 s_{11}^E}, \quad S_{eff} = 0.49wl, \quad S_{av} = 0.64wl, \quad n = 3.2\frac{wd_{31}t}{ls_{11}^E},$$

$$C_{el}^{S_1} = \varepsilon_{33}^T(1-k_{31}^2)\frac{4wl}{t}, \quad f_r = 0.45\frac{t}{l^2}\cdot\frac{1}{\sqrt{\rho s_{11}^E}}, \quad k_{eff}^2 = \frac{0.61k_{31}^2}{1-0.39]k_{31}^2}. \tag{2.130}$$

For the cantilever beam

$$K_{eqv}^E = 0.27\frac{wt^3}{l^3 s_{11}^E}, \quad S_{eff} = 0.24wl, \quad S_{av} = 0.40wl, \quad n = 0.66\frac{wd_{31}t}{ls_{11}^E},$$

$$C_{el}^{S_1} = \varepsilon_{33}^T(1-k_{31}^2)\frac{4wl}{t}, \quad f_r = 0.17\frac{t}{l^2}\cdot\frac{1}{\sqrt{\rho s_{11}^E}}, \quad k_{eff}^2 = \frac{0.42k_{31}^2}{1-0.58k_{31}^2}. \tag{2.131}$$

2.6. Flexural Type Transducers

The resonance frequencies of the transducers are determined by formula $f_r = 1/2\pi\sqrt{M_{eqv}C_m^E}$. Expressions for the effective coupling coefficients are calculated according to definition (2.93). The exact expression for the resonance mode of vibration for the simply supported beam is known as $\xi(x) = \xi_o \sin(\pi x / l)$. Values of the equivalent parameters calculated with using this expression differ from the approximate values obtained with mode of static deflection within 1-2%.

2.6.3 Circular Plate Transducer

As an example of a typical transducer for application in the receive mode consider the flexural circular bilaminar plate transducers shown in Figure 2.15. The plate is assumed to be thin compared with the radius $t \ll a$ and free of stress on the major surfaces. Thus, the stress in the axial direction throughout the thickness is neglected, and $T_3 = 0$. The electrodes are axially symmetric and the plate vibration as well.

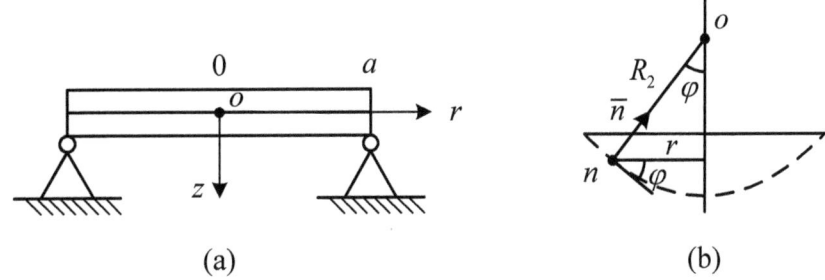

Figure 2.15: (a) Configuration of the circular plate transducer and (b) determination of the radius of curvature.

In the diametrical sections through the z-axis the radius of curvature is determined by the same expression as for the bent beam (see (2.114)), which was explained with reference to Figure 2.14. The second principal curvature is in the plane that is perpendicular to the rz-plane and goes through the normal of the bent plate (segment *no* in Figure 2.15 (b)). Point *o* on the intersection of normal with axis of symmetry of the plate is the center of curvature. The length of the segment *no* is the radius of curvature for the section of plate that goes in the circumferential direction, R_2. From geometry considerations $\varphi = -d\xi / dr$ and $r / R_2 = \sin\varphi \approx \varphi$ ($\varphi < \xi_o / a \ll 1$). Thus,

$$\frac{1}{R_1} = -\frac{d^2\xi}{dr^2}, \quad \frac{1}{R_2} = -\frac{1}{r}\frac{d\xi}{dr}, \qquad (2.132)$$

where the normal displacement is represented as

$$\xi(r/a) = \xi_0 \theta(r/a). \qquad (2.133)$$

By derivations analogous to those produced regarding Eq. (2.113)

$$S_1 = -z\xi_0 \cdot \frac{\partial^2 \theta}{\partial r^2}, \quad S_2 = -z\xi_0 \frac{1}{r}\frac{\partial \theta}{\partial r}. \qquad (2.134)$$

The piezoelectric equations suitable for this case are

$$S_1 = s_{11}^E T_1 + s_{12}^E T_2 + d_{31} E_3, \qquad (2.135)$$

$$S_2 = s_{12}^E T_1 + s_{22}^E T_2 + d_{31} E_3, \qquad (2.136)$$

$$D_3 = d_{31}(T_1 + T_2) + \varepsilon_{33}^T E_3. \qquad (2.137)$$

Upon substituting T_1 and T_2 from Eqs. (2.134) and (2.135) into Eq. (2.137), it will be obtained

$$D_3 = \frac{2d_{31}}{s_{11}^E + s_{12}^E}(S_1 + S_2) + \varepsilon_{33}^{S_{1,2}} \cdot E_3, \qquad (2.138)$$

where $\varepsilon_{33}^{S_{1,2}} = \varepsilon_{33}^T(1-k_p^2)$, $k_p^2 = 2d_{31}^2/\varepsilon_{33}^T(s_{11}^E + s_{12}^E)$. Thus, the capacitance of the transducer at parallel connection of the halves of the plate is

$$C_e^{S_{1,2}} = \varepsilon_{33}^T(1-k_p^2)4\pi a^2/t. \qquad (2.139)$$

It follows from Eqs. (2.135) and (2.136) that at $E_3 = 0$

$$T_1^E = \frac{s_{11}^E}{(s_{11}^E)^2 - (s_{12}^E)^2} S_1 - \frac{s_{12}^E}{(s_{11}^E)^2 - (s_{12}^E)^2} S_2, \qquad (2.140)$$

$$T_2^E = -\frac{s_{12}^E}{(s_{11}^E)^2 - (s_{12}^E)^2} S_1 + \frac{s_{11}^E}{(s_{11}^E)^2 - (s_{12}^E)^2} S_2. \qquad (2.141)$$

The potential energy of the plate vibration is

2.6. Flexural Type Transducers

$$W_{pot}^E = \frac{1}{2}\int_{\tilde{V}} \left(S_1 T_1^E + S_2 T_2^E\right) d\tilde{V} =$$

$$= \frac{1}{2}\xi_0^2 \frac{2\pi t^3}{12 s_{11}^E \left[1-(\sigma_1^E)^2\right]} \int_0^a \left[\left(\frac{\partial \theta}{\partial r}\right)^2 + 2\sigma_1^E \cdot \frac{1}{r}\frac{\partial \theta}{\partial r}\cdot\frac{\partial^2 \theta}{\partial r^2} + \left(\frac{1}{r}\frac{\partial \theta}{\partial r}\right)^2\right] r\, dr \qquad (2.142)$$

$$= \frac{1}{2}\xi_0^2 K_{eqv}^E,$$

where it is denoted $\sigma_1^E = -s_{12}^E/s_{11}^E$ as the analog of Poisson's ratio for piezoelectric ceramic material. The kinetic energy of the vibrating plate is

$$W_{kin} = \frac{1}{2}\dot\xi_0 \cdot 2\pi \rho t \int_0^a \theta^2(r/a) r\, dr = \frac{1}{2}\dot\xi_0 M_{eqv}. \qquad (2.143)$$

The equivalent mass will be denoted as $M_{eqv} = \rho t \cdot S_{eff}$, where

$$S_{eff} = 2\pi \int_0^a \theta^2(r/a) r\, dr \qquad (2.144)$$

is the effective surface area of the plate.

The electromechanical energy associated with vibration is

$$W_{em} = \frac{1}{2}\int_{\tilde V} D_3^E E_3(z) d\tilde V = \frac{1}{2}\int_{\tilde V} \frac{2 d_{31}}{s_{11}^E + s_{12}^E}(S_1+S_2) E_3 d\tilde V = \frac{1}{2}\xi_0 v n. \qquad (2.145)$$

The halves of the plate must be polarized and connected electrically as it is shown in Figure 2.13. Assuming that they are connected in parallel with $E_3 = 2V/t$, and after substituting expressions (2.134) for S_1 and S_2 into formula (2.145) will be obtained

$$W_{em} = \frac{1}{2}\xi_0 V(-)\frac{\pi d_{31} t}{s_{11}^E + s_{12}^E}\int_0^a \left(\frac{\partial^2 \theta}{\partial r^2} + \frac{1}{r}\frac{\partial \theta}{\partial r}\right) r\, dr = \frac{1}{2}\xi_0 V n, \qquad (2.146)$$

where from the electromechanical transformation coefficient is

$$n = -\pi \frac{d_{31} t a}{s_{11}^E + s_{12}^E}\frac{\partial \theta}{\partial r}\bigg|_{r=a}. \qquad (2.147)$$

Thus, the electromechanical transformation coefficient is proportional to the slope of the mode of vibration on the border of the electrode. The sign (–) is related to sign convention and in the context of calculating parameters of a single transducer it can be omitted as well as it was suggested regarding a beam. In the case that the circular plate piezoelement is used as accelerometer and its support vibrates with acceleration $\ddot{\xi}_s$, the equivalent force has to be determined by the expression analogous to (2.125)

$$F_{eqv} = \rho t \ddot{\xi}_s 2\pi \int_0^a \theta(r/a) dr = \rho t S_{av} \ddot{\xi}_s,$$ (2.148)

where

$$S_{av} = 2\pi \int_0^a \theta(r/a) dr$$ (2.149)

is the average surface of the plate. The mode of vibration of a circular plate is defined by the boundary conditions, which may in practice vary significantly depending on the transducer design. For the case of simply supported boundary (i.e., displacements and bending moment in the radial directions are zeros on the boundary), expression for the static deflection curve is

$$\theta(r/a) = \left(1 - \frac{r^2}{a^2}\right)\left(1 - \frac{1+\sigma}{5+\sigma}\frac{r^2}{a^2}\right) \approx \left(1 - \frac{r^2}{a^2}\right)\left(1 - \frac{r^2}{4a^2}\right).$$ (2.150)

These conditions may be closely achieved in the symmetrical double-sided transducer design shown in Figure 2.16. The plates assumed to be identical, therefore the design has the plane of symmetry that can be considered as absolutely rigid, and the boundaries of the plates don't move in the vertical direction when the plates vibrate.

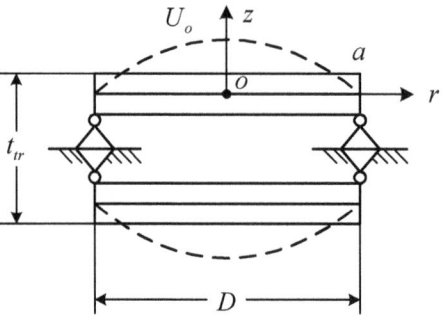

Figure 2.16: Symmetrical double-sided flexural transducer design.

A peculiarity of the above expressions for determining the equivalent parameters in the case of a simply supported circular plate by comparison with a beam is that they depend on the material parameter $\sigma = \sigma_1^E$, which affects the mode of vibration (2.150). Because of this the general formulas for the transducer equivalent parameters, strictly speaking, cannot be obtained

2.6. Flexural Type Transducers

in a closed form and calculation of the integrals involved must be fulfilled for each particular ceramic composition having different Poisson's ratio σ_1^E. However, the values of these integrals for all the modern ceramic materials do not deviate significantly from those obtained with an approximately average value of $\sigma_1^E = 0.3$. This will be shown in Section 4.5.7.1. If to neglect these small deviations, then after substituting the mode of vibration by formula (2.150) into the expressions for the equivalent parameters their following values will be obtained

$$K_{eqv}^E = \frac{23}{a^2} \frac{t^3}{12 s_{11}^E (1-\sigma_1^{E2})}, \quad S_{eff} = 0.29\pi a^2, \quad S_{avi} = 0.46\pi a^2,$$
$$n = 1.5\pi \frac{d_{31} t}{s_{11}^E + s_{12}^E}, \quad f_r = 0.24 \frac{t}{a^2} \frac{1}{\sqrt{s_{11}^E \rho}}.$$
(2.151)

The effective coupling coefficient of the circular plate are according to expressions (2.93)

$$\alpha_c = \frac{n^2}{K_{eqv}^E C_e^{S_{1,2}}} = 0.46 \frac{k_p^2}{1-k_p^2}, \quad k_{eff}^2 = \frac{\alpha_c}{\alpha_c + 1} = 0.46 \frac{k_p^2}{1 - 0.54 k_p^2}.$$
(2.152)

2.6.4 Acoustic Field Related Parameters of the Transducers of Flexural Type

Transducers of the flexural type are most commonly used in designs of the single hydrophones intended to react on the sound pressure or pressure gradient in acoustic field, and of the members of receiving arrays. Yet another field of their applications is in designs of the low frequency projectors: circular and rectangular benders. Configurations of the transducer designs for these applications are schematically sketched in Figure 2.16 and Figure 2.17. Detailed analyses of operating characteristics of the transducers will be done in Chapters 13 and 14. Information about the acoustic field related parameters of the transducers that may be needed for completing their considering on a basic level, namely, the radiation impedances and diffraction coefficients, is presented below.

The inherent property of transducers of the flexural type is the relatively small wave size of their mechanical systems. Among transducers of the flexural type the biggest wave size have benders that operate around their resonance frequencies. Using formula for the resonance frequency of the simply supported circular plate it can be shown that at the resonance frequency $D/\lambda \approx t/a$. For estimating the maximum wave size of a transducer, we assume that $t/a < 1/5$ (this is shown to be a reasonable relation in Chapter 9).

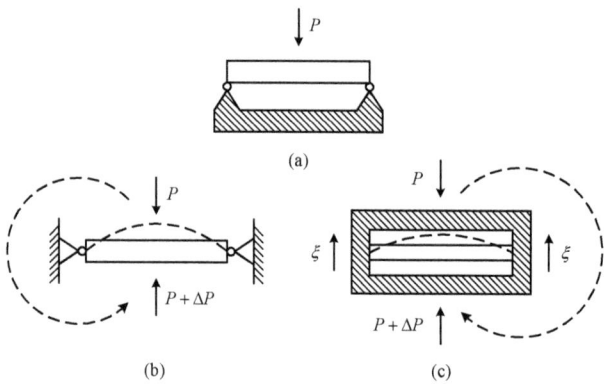

Figure 2.17: Configurations of the flexural transducer designs for different applications: (a) one-sided design of the benders and pressure hydrophones; (b) and (c) variants of designs of the pressure gradient hydrophones of the diffraction and motional type (definitions and details of calculation are given in Chapter 14).

Thus, it may be considered that $D/\lambda < 0.2$. The overall thickness, t_{tr}, of a single double-sided plate bender transducer (Figure 2.16) is usually about two thicknesses of the comprising plates, i. e., $t_{tr}/\lambda < 0.04$. As to the rectangular plate benders that are composed of beams, we will assume that a single transducer unit has approximately the same wave size as a circular plate having the same resonance frequency ($l = w = D$). Although this assumption is not quite rigorous, it allows fair enough comparison between benders of this kind.

One side of the double-sided symmetrical circular plate transducer can be considered as vibrating in the infinite rigid plane baffle due to symmetry. Therefore, the diffraction coefficient is $k_{dif} = 2$, and components of the radiation impedance per one side of the transducer vibrating in piston like mode would be determined by equations (2.104) through (2.107). Thus, for transducers of small wave size (at $ka < 0.6$) they would be

$$r_{ac} \approx \rho c \pi a^2 \frac{(ka)^2}{2} = 2\pi \rho c \frac{(\pi a^2)^2}{\lambda^2} \qquad (2.153)$$

within an accuracy of about 5%, and

$$x_{ac} \approx \rho c \pi a^2 \frac{8}{3\pi} ka = \omega \rho \frac{8}{3} a^3 = \omega M_{ac1} \qquad (2.154)$$

within an accuracy of about 10%.

2.6. Flexural Type Transducers

These formulas can be generalized for transducers of small wave size that have distribution of velocity on their surface by replacing surface area $S_\Sigma = \pi a^2$ by the average surface area S_{av} (because the volume velocity or "the source strength" $U_{\dot{v}} = S_{av}\dot{\xi}_o$ is what counts in terms of radiation of small sources). Thus, the expressions may be used

$$r_{ac1} \approx 2\pi\rho c \frac{(S_{av1})^2}{\lambda^2} \quad \text{and} \quad x_{ac1} \approx \rho c S_{av1} \frac{8}{3\pi} ka \qquad (2.155)$$

per one side of a double-sided transducer. The diffraction coefficient for this case remains the same, $k_{dif1} = 2$, being determined under assumption that surface of the transducer is clamped. These radiation impedances and diffraction coefficient may be used, when calculating magnitude of vibration $\dot{\xi}_o$ from the equivalent circuit derived for a single plate with simply supported boundary.

It is noteworthy that the same problem could be approached in the alternative way, if to consider the entire double-sided structure as a simple source vibrating in the free space. In this case in the general formula for radiation resistance of the simple source in the free space the total average surface (the total source strength) must be doubled, which will result in doubling the radiation resistance obtained for the previous case, as it follows from the manipulation

$$r_{ac} = \pi\rho c \frac{(S_{av})^2}{\lambda^2} = \pi\rho c \frac{(2S_{av1})^2}{\lambda^2} = 2r_{ac1}. \qquad (2.156)$$

The diffraction coefficient for this case will be found to be $k_{dif} = 1 = k_{dif1}/2$ as for a rigid body of small wave size in the free space. But now the equivalent electromechanical circuit, to which the equivalent acoustic parameters should be applied, must be derived for the entire double-sided structure. This will bring us to doubling the energies and corresponding equivalent parameters involved, and thus the final results of transducer calculation would be the same, as in the previous case. This must be taken into consideration in case the acoustic field related data are obtained from other sources.

The above estimations were made for a single bender transducer that can be considered approximately as a simple source. In practice, low frequency transducer designs often are composed of a number of elementary single sources, in our case from the single benders. The radiation impedance of a combination of several single bender transducers can be determined by considering the acoustic interaction between simple sources by formula

$$Z_{12} = r_{11}(\sin kd + j\cos kd)/kd, \qquad (2.157)$$

where Z_{12} is the mutual radiation impedance between two sources separated by distance d, and r_{11} is the radiation resistance of a single source that is given by expression (2.156). The origin of this formula and detailed analysis of possible results of interaction between transducers can be found in Ref. 10 and will be considered in Chapter 6.

The radiation problems that are associated with other applications of the flexural plate transducers that are schematically shown in Figure 2.17 (as one-sided and operating in the oscillating mode) were first considered in work [11]. The results related to our case of bodies having small wave size are as follows. For the piston like vibrating round disk radiating from one side at $ka < 0.6$

$$\alpha \approx \frac{(ka)^2}{4}, \quad \beta \approx \frac{2}{\pi}ka, \quad k_{dif} = 1. \qquad (2.158)$$

The corresponding radiation resistance for the plate vibrating with nonuniform distribution of velocity

$$r_{ac} = \pi(\rho c)_w \frac{(S_{av})^2}{\lambda^2}. \qquad (2.159)$$

For the oscillating disk

$$\alpha \approx \frac{8}{27\pi^2}(ka)^4, \quad \beta \approx \frac{4}{3\pi}ka. \qquad (2.160)$$

The radiation resistance of an oscillating body of small size is zero to the first approximation, because the total volume velocity (the source strength) is zero as velocities on the sides of a plate have different signs. And to the second approximation it is

$$r_{ac} = (\rho c)_w S_{av1} \frac{8}{27\pi^2}(ka)^4, \qquad (2.161)$$

where S_{av1} is the average surface per one side of the plate. The diffraction coefficient is

$$k_{dif} = j\frac{4}{3\pi}(ka)\cos\theta, \qquad (2.162)$$

and the average surface of the plate that must be taken in calculation of the equivalent force is $2S_{av1}$. Thus,

$$F_{eqv} = 2S_{av1}k_{dif}P_o. \qquad (2.163)$$

In conclusion it has to be noted that transducers of all the types considered in this chapter have to be treated as the systems with multiple degrees of freedom under more general assumptions regarding dimensions of the piezoelements, modes of their polarization, configuration of the electrodes, acoustic loading, and with effects of coupled vibrations in in their mechanical systems. The general treatment of these transducers will be performed in Chapters 7-10.

2.7 References

1. S. P. Timoshenko, *Vibration Problems in Engineering*, 2nd Ed. (Van Nostrand, New York, 1937).
2. P. M. Morse, *Vibration and Sound*, 2nd Ed. (McGraw-Hill, New York, 1948).
3. P. M. Morse and K. U. Ingard, *Theoretical Acoustics* (McGraw-Hill, New York, 1968).
4. *Handbook of Mathematical Functions*, edited by M. Abramowitz and I. A. Stegun (National Bureau of Standards Applied Mathematical Series, Washington, DC, 1964).
5. D. H. Robey, "On the radiation impedance of an array of finite cylinders," J. Acoust. Soc. Am. **27**, 706-710 (1955).
6. B. S. Aronov, "Energy analysis of a piezoelectric body under nonuniform deformation," J. Acoust. Soc. Am. **113**(5), 2638-2646 (2003).
7. W. G. Cady, *Piezoelectricity* (McGraw-Hill, New York, 1946).
8. L. E. Kinsler, A. R. Frey, A. B. Coppens, and J. V. Sanders, *Fundamentals of Acoustics*, 4th Ed. (John Wiley & Sons, New York, 2000).
9. S. P. Timoshenko, *Strength of Materials*, Part I, Elementary Theory and Problems, 2nd Ed. (Van Nostrand, New York, 1940).
10. R. L. Pritchard, "Mutual acoustic impedance between radiators in an infinite rigid plane," J. Acoust. Soc. Am. **32**(6), 730-737 (1960).
11. L. Y. Gutin, "A sound field of the piston-like projectors", Zhurnal Tekhnicheskoi Fiziki, Vol. 7, No. 10, 1937. Selected works in Shipbuilding (Sudostroenie, Leningrad, 1977), p. 95 (in Russian).

CHAPTER 3

TRANSDUCER PERFORMANCE ANALYSIS

3.1 Operation in Transmit Mode

Transducer operating in the transmit mode must be considered as a part of the transmit channel, that is as a single transducer or a member of a group of transducers driven by a common power amplifier. The goal to be achieved by a transmit channel is to provide a certain acoustic intensity in the direction of the acoustic axis of the projector, $I(0)$, and required spatial distribution of sound pressure described by the directional factor $H(\theta,\phi)$. By definition

$$I(0) = |P(0)_{1m}|^2 / (\rho c)_w, \qquad (3.1)$$

where $P(0)_{1m}$ is the effective sound pressure (rms) on the acoustic axis in the far field referenced to 1 meter from the acoustic center of the transducer, and

$$H(\theta,\phi) = P(\theta,\phi) / P(0,0). \qquad (3.2)$$

The units to measure the intensity and sound pressure are W/m² and Pa. As the magnitudes of the acoustic quantities may change in a very wide range, it is common to characterize them in the logarithmic to the base 10 scale relative to some reference levels of the quantities. Namely, for the Intensity Level (*IL*) and for the Sound Pressure Level (*SPL*)

$$IL = 10\log(I / I_{ref}), \qquad (3.3)$$

$$SPL = 20\log(P / P_{ref}). \qquad (3.4)$$

It is customary in underwater acoustics to use $P_{ref} = 1$ µPa and $I_{ref} = 6.76 \cdot 10^{-19}$ W/m² ($I_{ref} = P_{ref}^2 / (\rho c)_w$). Thus,

$$SPL = 20\log[P(0)] \text{ re } 1 \text{ µPa at } 1 \text{ m}. \qquad (3.5)$$

The requirements for a transmit channel must be met in a prescribed frequency band and under the condition of a long-term reliable operation at specified environmental conditions.

Our treatment of the transmit channel will be concentrated mainly on the part of the problem related to the transducer as an electromechanical device. In this section characteristics of a transducer will be considered that must be known for fulfilling the requirements for the transmit channel, given that they can be fully described by the equivalent electromechanical circuit of the transducer with acoustic load known. An assumption is that the acoustic load can be changed by means of the mechanical-acoustic matching, and the electrical input impedance of the transducer subject to tuning by using the additional inductances in order to facilitate conditions for matching with internal impedance of power amplifier. The transducers will be considered as having one mechanical degree of freedom. This assumption is applicable for the most of projectors operating in the vicinity of their resonance frequency.

3.1.1 Transducer Input Impedance and Tuning Conditions

A significant part of the problem of generating acoustic energy in a broad frequency band is that of delivering electrical power to a highly reactive and frequency dependent load, which the piezoceramic transducer presents. Severity of the driving source problems depends on how the input impedance of a projector and internal impedance of the source are matched. The overall efficiency of conversion from electric power to acoustic power produced by a transmit channel is determined by the combined losses in the source of energy, in the matching network, and in the transducer. Therefore, the projector must be appropriately designed for matching, and developers of the power supply should be fully aware of peculiarities of the transducer input in order to rationally design the matching network. (Obviously, the best result can be achieved in the case that developing the entire transmit channel is in the same hands.) The goal of transducer input impedance analysis is in exploring conditions for compensating its reactive component (conditions for tuning) in as broad frequency band as possible and thus in presenting transducer as predominantly active load for power amplifier.

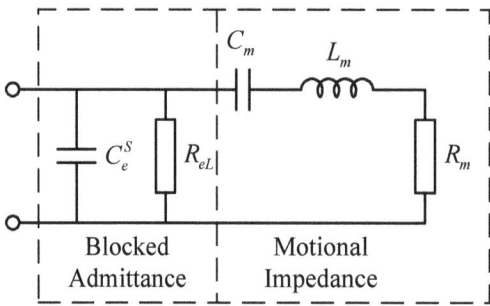

Figure 3.1: Equivalent circuits of a transducer recalculated to its electrical side.

3.1. Operation in Transmit Mode

The general expressions for the electrical input impedance of a transducer, Z_{in}, and input impedance of its mechanical system, Z_m^E, are given by formulas (1.80) and (1.75), and are illustrated by the equivalent circuit in Figure 1.17. The equivalent circuit of a transducer after recalculating the mechanical contour of the equivalent circuit into the electrical side is given in Figure 3.1, where

$$C_m = C_m^E n^2, \quad L_m = (M_{eqv} + m_{ac})/n^2, \quad R_m = (r_{mL} + r_{ac})/n^2. \tag{3.6}$$

In the most cases the projectors operate in the frequency region around the resonance frequencies of their mechanical systems. The resonance frequency in water, ω_{rw}, may be found from the equation $\text{Im}\{Z_m^E\} = 0$, or

$$\frac{\omega_{rw}}{\omega_{ra}}\left(1 + \frac{x_{ac}}{\omega_{rw} M_{eqv}}\right) - \frac{\omega_{ra}}{\omega_{rw}} = 0. \tag{3.7}$$

Usually for the typical underwater projectors $x_{ac}/(\omega_{ra} M_{eqv}) \ll 1$ in vicinity of their resonance frequencies (see the examples in Table 3.1), and without a significant error ω_{rw} in the parenthesis in equation (3.7) may be replaced by ω_{ra}. Under this assumption it follows from Eq. (3.7) that

$$\omega_{rw} \approx \frac{\omega_{ra}}{\sqrt{1 + x_{ac}/\omega_{ra} M_{eqv}}} \tag{3.8}$$

and expression (1.75) for Z_m^E becomes

$$Z_m^E = (r_{ac} + r_{mL})\left[1 + j\frac{\omega_{rw} M_{eqv}}{r_{ac} + r_{mL}}\left(\frac{\omega}{\omega_{rw}} - \frac{\omega_{rw}}{\omega}\right)\right]. \tag{3.9}$$

As $\omega_{rw} \approx \omega_{ra}$, and in a narrow frequency band around the resonance frequency it can be assumed that $r_{ac} + r_{mL}$ is constant, we can consider that quantity

$$\frac{\omega_{rw} M_{eqv}}{r_{ac} + r_{mL}} \approx \frac{\omega_{ra} M_{eqv}}{r_{ac} + r_{mL}} = Q_{mw}, \tag{3.10}$$

which we define as the mechanical quality factor of the transducer in water, remains approximately constant in this frequency band. Under this assumption expression (3.9) becomes

$$Z_M^E = (r_{ac} + r_{mL})\left[1 + jQ_{mw}\left(\frac{f}{f_{rw}} - \frac{f_{rw}}{f}\right)\right]. \tag{3.11}$$

The frequency f may be represented as

$$f = f_{rw} \pm \Delta f, \qquad (3.12)$$

where Δf is the deviation of frequency from its value at resonance. Substituting (3.12) for the term in parenthesis in Eq. (3.11), after some manipulations we arrive at

$$\frac{f}{f_{rw}} - \frac{f_{rw}}{f} \doteq \pm \frac{2\Delta f}{f_{rw}}\left[1 \mp \frac{\Delta f}{2f_{rw}} \pm 2\left(\frac{\Delta f}{2f_{rw}}\right)^2 \mp \ldots\right]. \qquad (3.13)$$

Table 3.1: Acoustic loads and their effect on the resonance frequencies for different transducer types.

	f_{ra}	M_{eqv}	$\alpha = \dfrac{r_{ac}}{(\rho c)_w S_r}$	$\beta = \dfrac{x_{ac}}{(\rho c)_w S_r}$	$Q_{mw} = \dfrac{\omega_{ra} M_{eqv}}{r_{ac}}$	$\dfrac{x_{ac}}{\omega_{ra} M_{eqv}}$	$\dfrac{f_{rw}}{f_{ra}}$
Sphere (Sect. 2.2)	$\dfrac{1.7 c_c}{2\pi a}$	$\rho t S_r$, $S_r = 4\pi a^2$	0.93	0.26	$30\dfrac{t}{a}$	$0.01\dfrac{a}{t}$	$\left(\sqrt{1+0.01\dfrac{a}{t}}\right)^{-1}$
	For PZT-4, $ka = 3.7$				6.0 [2]	0.05 [2]	0.98 [2]
Cylinder[1] (Sect. 2.3)	$\dfrac{c_c}{2\pi a}$	$\rho t S_r$, $S_r = 2\pi ah$	0.92	0.20	$18\dfrac{t}{a}$	$0.01\dfrac{a}{t}$	$\left(\sqrt{1+0.01\dfrac{a}{t}}\right)^{-1}$
	For PZT-4, $ka = 2.2$				3.6 [2]	0.05 [2]	0.98 [2]
Uniform Bar [3] (Sect. 2.4)	$\dfrac{c_c}{2l}$	$\dfrac{\rho}{2} l S_r$	In a large array				
			1.0	0.0	25	0	1

[1] Per unit height of a ring operating in a long column like transducer.
[2] Value for $a/t = 5$.
[3] Single bar transducer operating in a plane array of a big size vibrating in phase. Ring and bar can employ transverse or longitudinal piezoeffect. In the table the transverse piezoeffect is assumed to be used. Switching to the longitudinal piezoeffect would not change results in principle.

If to assume that

$$\Delta f / f_{rw} \leq 0.2, \qquad (3.14)$$

which is appropriate for most of the practical cases, then within 10% accuracy

3.1. Operation in Transmit Mode

$$\frac{f}{f_{rw}} - \frac{f_{rw}}{f} \doteq \pm \frac{2\Delta f}{f_{rw}} = \pm \Omega. \quad (3.15)$$

Notation Ω is for the relative deviation of frequency from its value at resonance. Finally, we arrive at expression

$$Z_m^E \approx (r_{ac} + r_{mL})(1 + j\Omega Q_{mw}) \quad (3.16)$$

for the band of $|\Omega| \leq 0.4$. The mechanical impedance may be presented as

$$Z_m^E = |Z_m^E| e^{j \arg Z_m^E}, \quad (3.17)$$

where

$$|Z_m^E| = (r_{ac} + r_{mL})\sqrt{1 + (\Omega Q_{mw})^2}, \quad \arg Z_m^E = \arctan \Omega Q_{mw}. \quad (3.18)$$

Quantitative estimations of the parameters Q_{mw} and f_{rw} vs. f_{ra} related to some typical transducer mechanical systems with acoustic load are presented in Table 3.1.

The equivalent circuit of the transducer electrical side that includes the motional impedance (Figure 3.1) can be further transformed into the circuit with parallel representation of the motional impedance that is shown in Figure 3.2. Parameters of the circuit in Figure 3.2 (a) are related to the original mechanical parameters of a transducer by formulas (3.6) and by the following relations

$$R_p = \frac{X_m^2 + R_m^2}{R_m} = \frac{1}{G_{mw}}, \quad (3.19)$$

$$\Delta C_p = -\frac{X_m}{\omega(X_m^2 + R_m^2)}, \quad (3.20)$$

where

$$X_m = \frac{1}{\omega C_m}\left(\frac{\omega^2}{\omega_r^2} - 1\right) = -\frac{1}{\omega C_m^E n^2} \Omega. \quad (3.21)$$

Upon substituting the expression X_m into the Eq. (3.19) and (3.20) we obtain

$$G_{mw}(\Omega) = \frac{1}{R_p} = G_{mw}(0)\left(\frac{1}{1 + \Omega^2 Q_{mw}^2}\right), \quad (3.22)$$

$$\Delta C_p = C_m^2 n^2 \left(\frac{\Omega Q_{mw}^2}{1+\Omega^2 Q_{mw}^2} \right). \qquad (3.23)$$

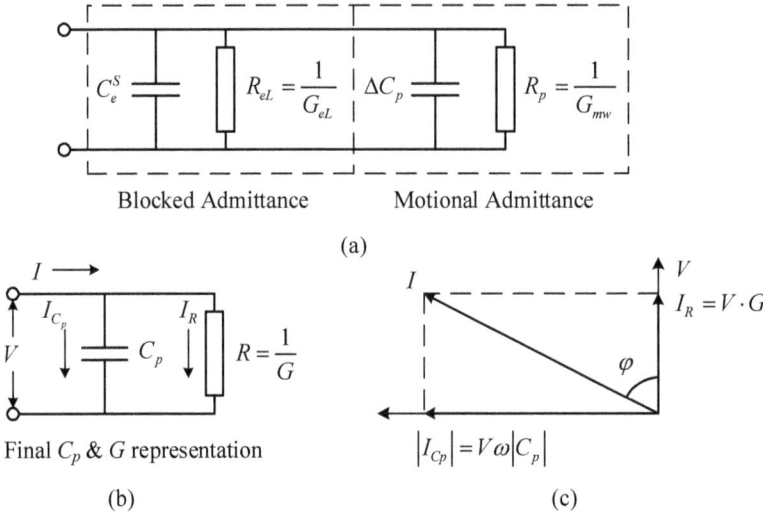

Figure 3.2: Electrical circuits of the projector input: (a) as the combination of the blocked and motional admittances, (b) final representation as measured by an impedance analyzer, (c) V, I_G vector diagram.

The functions ΔC_p and G_{mw} vs. frequency deviation Ω are qualitatively illustrated in Figure 3.3.

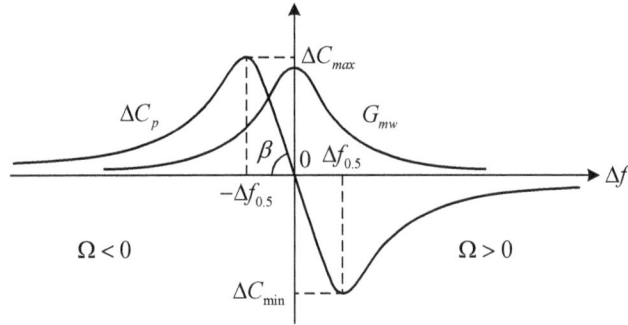

Figure 3.3: The functions ΔC_p and G_{mw} vs. frequency deviation from the resonance frequency.

In order to determine the extremes of ΔC_p and its slope at $\Delta f = 0$, we differentiate the right-hand side of Eq. (3.20) with respect to Δf

3.1. Operation in Transmit Mode

$$(\Delta C_p)'_{\Delta f} = (\Delta C_p)'_{\Omega} \frac{2}{f_r} = -\frac{2C_m^E n^2}{f_r} \left[\frac{Q_{mw}^2(1+\Omega^2 Q_{mw}^2) - 2\Omega^2 Q_{mw}^4}{(1+\Omega^2 Q_{mw}^2)^2} \right]. \quad (3.24)$$

Thus, at $\Delta f = 0$

$$(\Delta C_p)'_{\Delta f=0} = \tan \beta = -\frac{2}{f_r} C_m^E n^2 Q_{mw}^2. \quad (3.25)$$

From the equation $(\Delta C_p)' = 0$ it follows that $\Omega^2 Q_{mw}^2 = 1$. Next, we will determine the frequency deviations corresponding to the extreme values of ΔC_p as

$$\Delta f_m = \pm \frac{f_r}{2 Q_{mw}}. \quad (3.26)$$

After substituting $\Omega^2 Q_{mw}^2 = 1$ into Eq. (3.22), we will find that $G_{mw}(f_m) = G_{mw}(0)/2$, which means that at the deviation Δf_m, the energy of the mechanical vibration drops to 0.5 of its value at the resonance frequency. Therefore, this deviation can be denoted as $\Delta f_m = \pm \Delta f_{0.5}$. Upon substituting $\Omega Q_{mw} = \pm 1$ into Eq. (3.23), we arrive at

$$\Delta C_{max} = -\Delta C_{min} = C_m^E n^2 \frac{Q_{mw}^2}{2}. \quad (3.27)$$

Taking into account expressions (3.22) and (3.23) we finally represent the parameters C_p and G of electric circuit of the input admittances shown in Figure 3.2 (b) as

$$C_p = C_e^S + \Delta C_p = C_e^S \left[1 - \frac{C_m^E n^2}{C_e^S} \left(\frac{\Omega Q_{mw}^2}{1+\Omega^2 Q_{mw}^2} \right) \right]. \quad (3.28)$$

Remembering (see (2.93)) that

$$\frac{C_m^E n^2}{C_e^S} = \alpha_c = \frac{k_{eff}^2}{1-k_{eff}^2}, \quad (3.29)$$

and representing

$$\frac{n^2 R_{eL}}{r_{mL} + r_{ac}} = \frac{n^2 \omega C_m^E Q_{mw}}{\omega C_e^S / Q_e} = \left(\frac{k_{eff}^2}{1-k_{eff}^2} \right) Q_e Q_{mw} \quad (3.30)$$

the equations for C_p and G can be transformed to

$$C_p = C_e^S \left[1 - \frac{k_{eff}^2}{1-k_{eff}^2} \left(\frac{\Omega Q_{mw}^2}{1+\Omega^2 Q_{mw}^2} \right) \right] \quad (3.31)$$

and

$$G = \frac{1}{R_{eL}}\left[1 + \frac{k_{eff}^2}{1-k_{eff}^2}\left(\frac{Q_e Q_{mw}^2}{1+\Omega^2 Q_{mw}^2}\right)\right]. \quad (3.32)$$

(Note that for completed transducer design $Q_e \neq (1/\tan\delta_m)$ due to possible stray losses) Thus, the input impedance of a projector is represented by the parallel connection of frequency dependent "capacitance," $C_p(\omega)$, and resistance, $R_p(\omega)$. (Note that depending on the combination of k_{eff} and Q_{mw} in some frequency band function $C_p(\omega)$ may become negative, and in this case C_p can only conditionally be called "capacitance"). In order to characterize the input impedance as electric load to driver amplifier, we introduce the electric quality, Q_{ew}, and power factor, $\cos\varphi$, of the loaded projector by the following relations

$$Q_{ew} = \omega |C_p(\omega)| \cdot R_p(\omega) \quad (3.33)$$

and

$$\cos\varphi = \frac{I_R}{I} = \frac{I_R}{\sqrt{I_R^2 + I_{C_p}^2}} = \frac{1}{\sqrt{1+Q_{ew}^2}}. \quad (3.34)$$

The relation between voltage across the projector, V_{tr}, and input current, I_{tr}, is

$$|I_{tr}| = |V_{tr}|\frac{1}{R_p \cos\varphi}. \quad (3.35)$$

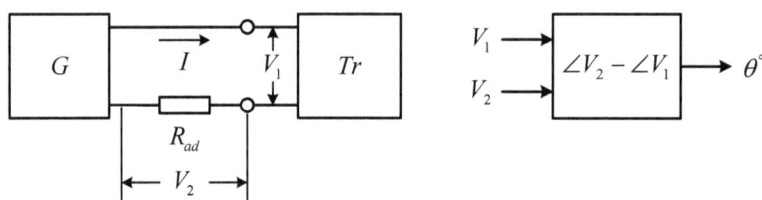

Figure 3.4: Experimental set up for determining the transducer admittance: G – function generator, Tr – transducer, θ° - phase meter.

The input admittance of a transducer

$$Y_{in} = 1/Z_{in} = I/V, \quad Y_{in} = |Y_{in}|e^{j\theta^\circ} \quad (3.36)$$

3.1. Operation in Transmit Mode

can be experimentally determined with aid of the set up presented in Figure 3.4, which allows to get in principle the same results in terms of the input admittance, as can be produced by an impedance analyzer, a device that is not always available because of its high cost. The additional resistance R_{add} should be chosen in such a way that

$$R_{ad} \ll |Z_{in}| \qquad (3.37)$$

in all the frequency band of measurements. Under this condition

$$V_2 / R_{ad} = I = V_1 Y_{in}, \qquad (3.38)$$

where from

$$|Y_{in}| = |V_2|/R_{ad}|V_1| \text{ and } \theta° = \arg\{Y_{in}\} = \arg\{V_2\} - \arg\{V_1\}. \qquad (3.39)$$

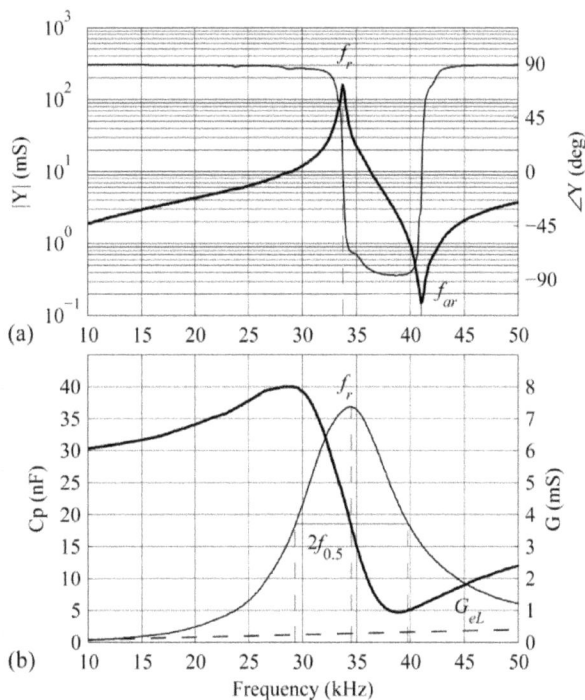

Figure 3.5: Components of the input admittance of the spherical transducer (PZT-4, $2a = 1.9$ in, $t = 0.1$ in): (a) modulus and phase of the admittance in air, (b) parallel capacitance C_p and conductance G in water.

The dependences of components of the admittance on frequency qualitatively look as shown with plots in Figure 3.5. They were measured with example of a spherical transducer in air and in water. Dependence from frequency of conductance G_{eL} that corresponds to the electrical losses, which is shown in Figure 3.5 by the dashed line, should be linear according the assumption made by relation (1.39) (at least within limits, in which factor $\Delta_{eL} = \tan\delta_e$ remains independent of frequency). Next to zero point at this line have to be chosen at a frequency that is much lower than resonance frequency of the transducer in order to minimize a contribution of the mechanical losses.

At first consider the input admittance of an unloaded transducer (at $Z_{ac} = 0$). This mode of operation can be used for experimental determining of unknown parameters of the transducer impedance such as resistances of the mechanical and electrical losses, for measuring important transducer characteristics such as the resonance frequency and effective coupling coefficient, for control of these parameters in course of transducers fabrication. Besides this mode of operation is typical of transducers used as resonators that represent two-terminal networks. The measurements performed in air result in the most accurate data regarding the characteristics that are inherent in the transducer piezoelement properties.

The resonance frequencies of the electrical contour in Figure 3.1 correspond to the conditions that $\arg\{Y\} = 0$. They are the frequency of the parallel resonance that coincides with the resonance frequency of transducer mechanical system,

$$f_p = f_r = 1/2\pi\sqrt{L_m C_m} = 1/2\pi\sqrt{M_{eqv} C_m^E}, \qquad (3.40)$$

and frequency of the series resonance that is called the antiresonance frequency of the transducer,

$$f_s = f_{ar} = f_r [1 + C_m^E n^2 / C_e^S]^{1/2}. \qquad (3.41)$$

Combination of the equivalent parameters in brackets was denoted as α_c in the expression (3.29) for the effective coupling coefficient. After manipulations that involve this expression, it will be obtained from formula (3.41) that

$$k_{eff}^2 = 1 - (f_r / f_{ar})^2. \qquad (3.42)$$

As it follows from Eq. (3.22), without the acoustical load

$$\frac{G_m(\Omega)}{G_m(0)} = \left(\frac{1}{1+\Omega^2 Q_m^2}\right). \qquad (3.43)$$

3.1. Operation in Transmit Mode

Thus, at $G_m(\Omega)/G_m(0) = 1/2$, $\Omega Q_m = 1$ and

$$Q_m = f_r / 2\Delta f_{0.5}, \quad (3.44)$$

where $\Delta f_{0.5}$ is deviation of frequency at which mechanical conductance reaches 0.5 of its value at resonance frequency. According to notation (3.10) $Q_m = \omega_r M_{eqv} / r_{mL} = 1/\omega_r C_m r_{mL}$, i. e., the mechanical quality factor.

Measuring parameters of transducer in water allows determining values of components of the radiation impedance. From Eq. (3.8) follows that

$$\frac{x_{ac}}{\omega_{ra} M_{eqv}} = \frac{f_{ra} - f_{rw}}{f_{rw}}. \quad (3.45)$$

Difference between f_{ra} and f_{rw} in the denominator can be neglected, and we arrive at the approximate formula for the reactive component of the radiation impedance

$$x_{ac} \approx 2\pi M_{eqv}(f_{ra} - f_{rw}). \quad (3.46)$$

Peculiarity exists when dealing with a transducer encapsulated in polyurethane (PU) or a like material. The mass of PU is added to equivalent mass of the transducer when resonance frequency is measured in air. This lowers the original resonance frequency of a bare piezoelement. When measured in water, the layer of PU becomes transparent, and the additional mass disappears. This must be taken into consideration in calculating the acoustic reactance from results of experimenting.

At the resonance frequencies in air and in water from Eq. (3.16) follows that $R_p = R_m$, i.e.,

$$G_{mw}(0) = \frac{n^2}{r_{mL} + r_{ac}} \quad \text{and} \quad G_{ma}(0) = \frac{n^2}{r_{mL}}, \quad (3.47)$$

when measured in water and in air, respectively. Thus,

$$\frac{r_{ac}}{n^2} = \frac{1}{G_{mw}(0)} - \frac{1}{G_{ma}(0)}. \quad (3.48)$$

Given that all the electromechanical transducer parameters are known the radiation resistance can be determined.

3.1.1.1 On the Tuning Conditions

Tuning of input impedance is achieved by inserting an inductor in parallel or in series with the projector input as shown in Figure 3.6. The inductances L_p, L_s and the respective

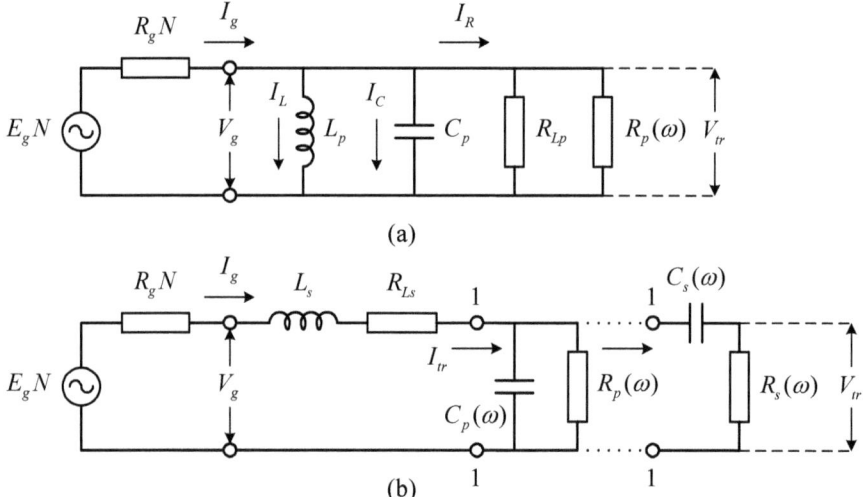

Figure 3.6: Electrical circuit of the transmit channel: (a) with parallel tuning, (b) with series tuning.

resistances R_{Lp} and R_{Ls}, which are accountable for the energy losses in the inductors, are related through the quality factor of the inductor Q_L. Using formulas for conversion from the parallel to series impedance representation

$$R_s = R_p \cdot \frac{1}{1+\left(R_p/X_p\right)^2}, \quad X_s = X_p \cdot \frac{1}{1+\left(X_p/R_p\right)^2} \tag{3.49}$$

and the definition for the quality factor $Q = \mathrm{Im}\{\vec{W}\}/\mathrm{Re}\{\vec{W}\}$ we obtain for the same inductor in different representations

$$Q_L = \frac{\omega L_s}{R_{Ls}} = \frac{R_{Lp}}{\omega L_p} \quad \text{and} \quad L_s = L_p \frac{Q_L^2}{1+Q_L^2}. \tag{3.50}$$

The quality factor, Q_L, can be considered as frequency independent in a given frequency band. When operating in the frequency band around the resonance frequency usually (but not exclusively) the frequency of mechanical resonance is chosen as the frequency of exact tuning. In the rare cases of operating in the frequency band well below the resonance the exact tuning is fulfilled at the desired operating frequency. In this section we will assume that the frequency of exact tuning is denoted f_t.

The main goal of tuning is to reduce the reactive load as seen by the electric power source, or to reduce the total power $|\overline{W_{el}}| = |V_g||I_g|$ delivered to the tuning circuit in comparison with the total power $|\overline{W_{tr}}| = |V_{tr}||I_{tr}|$ required for the operation of the untuned projector.

In the case of the parallel circuit $V_g = V_{tr}$ and at the frequency of exact tuning.

$$|I_g| = |I_{tr}| \cdot \frac{1}{\sqrt{1+Q_{ew}^2}} \left(1 + \frac{Q_{ew}}{Q_L}\right), \qquad (3.51)$$

as it follows from analysis of the circuit in Figure 3.6 (a) and the definition for Q_{ew} by formula (3.33). In the case that $Q_L \gg Q_{ew}$, which is desirable in order to avoid noticeable losses of energy in the tuning circuits, and taking into account formula (3.34) for the power factor of the transducer, the current I_g can be represented as

$$|I_g| \approx |I_{tr}| \cos \varphi_{tr} \qquad (3.52)$$

and

$$|V_g||I_g| \approx |V_{tr}||I_{tr}| \cos \varphi_{tr}. \qquad (3.53)$$

It must be noted that both Q_{ew} and Q_L may become of the same order of magnitude, when operating in a broad frequency band around the resonance frequency of a projector and especially at frequencies much below the resonance. In this case the factor in parenthesis of formula (3.51) cannot be neglected, and the gain in power supplied that is expressed by expression (3.53) will be less impressive.

In the case of the series tuning

$$I_g = I_{tr} \text{ and } |V_g| = |V_{tr}| \cdot \frac{1}{\sqrt{1+Q_{ew}^2}} \left(1 + \frac{Q_{ew}}{Q_L}\right), \qquad (3.54)$$

as it follows from analysis of the circuit in Figure 3.6 (b). Thus,

$$|V_g| \approx |V_{tr}| \cos \varphi_{tr} \text{ and } |V_g||I_g| \approx |V_{tr}||I_{tr}| \cos \varphi_{tr} \qquad (3.55)$$

under the same assumption regarding value of Q_L, i.e., the same result is obtained as in the case of the parallel tuning. The difference between these circuits is that in the case of the parallel tuning the equivalent generator that represents the driving amplifier with the matching transformer has to be capable of producing a high driving voltage and a moderate output current, while in the case of the series tuning it has to be capable of providing a larger current under the smaller output voltage. In terms of requirements for the inductors, the operating conditions, which they face in the parallel and series circuits, are approximately the same. Thus, in the parallel circuit at the frequency of exact tuning

$$V_L = V_{tr} \text{ and } |I_L| = |I_{tr}| \cdot \frac{Q_{ew}}{\sqrt{1+Q_{ew}^2}} \cdot \frac{\sqrt{1+Q_L^2}}{Q_L}, \qquad (3.56)$$

and in the series circuit

$$I_L = I_{tr} \text{ and } |V_L| = |V_{tr}| \cdot \frac{Q_{ew}}{\sqrt{1+Q_{ew}^2}} \cdot \frac{\sqrt{1+Q_L^2}}{Q_L}. \qquad (3.57)$$

Taking into account that $Q_L \gg 1$ and usually $Q_{ew} \gg 1$, it can be concluded that the voltage across the inductor may be slightly smaller in the case of the series circuit, though basically the tuning inductors in the transmit channel operate under the same high voltage as the projectors do, and the practical realization of an inductor with desirably high Q_L that operates at high voltage may be a challenging.

Conditions of operating the tuning circuits depend essentially on the behavior of $C_p(\omega)$ and $R_p(\omega)$ parameters of the transducer input impedance. Therefore, the results of application of formulas describing effects of tuning will be different for the frequency bands A and B in Figure 3.7, where the plots of $C_p(\omega)$ and $G = 1/R_p(\omega)$ vs. frequency are qualitatively shown. In the band B, where $C_p(\omega)$ changes slowly, the exact tuning may take place at a single frequency, f_t, according to the formula

$$\frac{1}{\omega_t^2 L_p} = C_p(\omega_t). \qquad (3.58)$$

Quite a different situation takes place in the vicinity of the resonance frequency (in the band A), where the capacitance $C_p(\omega)$ may change rapidly depending on the mechanical quality factor of acoustically loaded projector and its effective coupling coefficient. Usually, the intended frequency for exact tuning in this frequency band is the resonance frequency of the projector, thus, it should be

3.1. Operation in Transmit Mode

$$L_p = \frac{1}{\omega_r^2 C_p(\omega_r)} = \frac{1}{\omega_r^2 C_e^S}. \tag{3.59}$$

At certain conditions the functions

$$F(\omega) = \frac{1}{\omega^2 L_p} = \frac{f_r^2}{f^2} C_e^S \tag{3.60}$$

and $C_p(\omega)$ may intersect in two more points, f_{t1} and f_{t2}, in addition to the point f_r, as is shown in Figure 3.7.

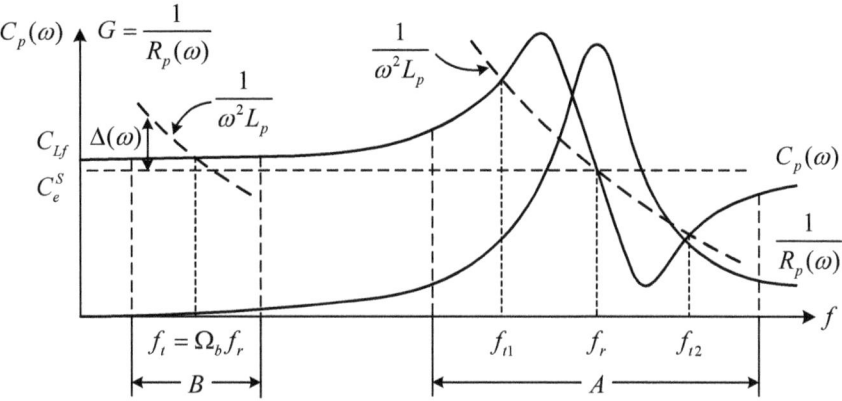

Figure 3.7: Qualitative plot of dependencies of C_p and G components of the input admittance.

Thus, the exact tuning (with power factor $\cos\varphi = 1$) can be achieved at three frequencies. Between these frequencies the reactance remains partly not compensated ($\cos\varphi < 1$). The non-compensated part of the parallel reactive element (capacitance or inductance depending on the frequency deviation from the frequency of exact tuning) at the particular frequencies is characterized by the segments $\pm\Delta(\omega)$ shown in the Figure. The vector diagram for currents through the transducer input admittance at these frequencies is shown in Figure 3.8, where

$$\Delta(\omega)\cdot\omega = \left|\mathrm{Im}\{\Delta Y(\omega)\}\right|, \tag{3.61}$$

and the power factor at these points is

$$\cos\varphi = \frac{G}{\sqrt{G(\omega)^2 + [\Delta(\omega)\cdot\omega]^2}}. \tag{3.62}$$

Consider the tuning situation in the region A in more detail. The region A of change of $C_p(f)$ and function $F(f)$ determined by formula (3.60) is shown in Figure 3.9.

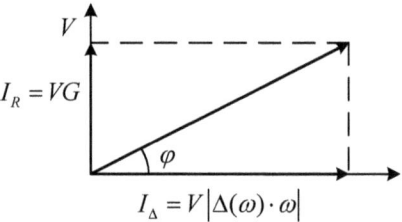

Figure 3.8: Vector diagram for currents through the transducer input admittance.

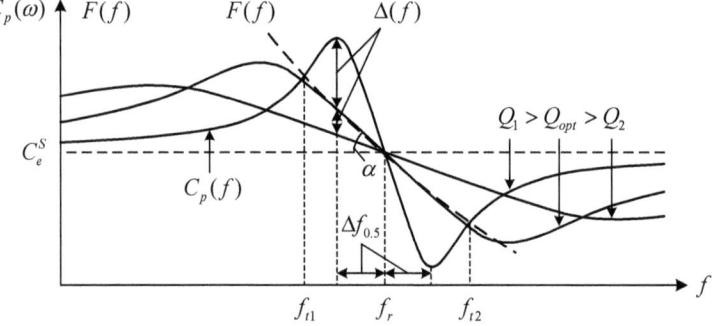

Figure 3.9: On the tuning in the region A: $F(f) = \left(f_r^2 / f^2\right) C_e^S$.

According to formula (3.25) the slope of the function $C_p(f)$ at $f = f_r$ is determined by the expression

$$\tan \alpha = -\frac{2}{f_r} C_m^E n^2 Q_{mw}^2 = -\frac{2}{f_r}\left(\frac{k_{eff}^2}{1-k_{eff}^2}\right) Q_{mw}^2 C_e^S. \tag{3.63}$$

We denote

$$\left(\frac{k_{eff}^2}{1-k_{eff}^2}\right) Q_{mw}^2 = Q. \tag{3.64}$$

The factor Q is related to the mechanical quality factor of the loaded transducer, Q_{mw}, and to the effective coupling coefficient of the transducer, k_{eff}, and characterizes the slope of the function $C_p(f)$ at the resonance frequency of the transducer. Behavior of $C_p(f)$ for transducer having different values of Q is qualitatively illustrated in Figure 3.9. The function $F(f)$ expressed by formula (3.60) will have minimal deviations from $C_p(f)$ in the case that the slopes of both functions at $f_t = f_r$ are equal, i.e., under the condition

$$F'(f)\big|_{f=f_r} = -\frac{2}{f_r} \cdot C_e^S = -\frac{2}{f_r} \cdot Q \cdot C_e^S. \tag{3.65}$$

3.1. Operation in Transmit Mode

The right hand part of Eq. (3.65) represents expression (3.63) for $\tan \alpha$ after factor Q is substituted. Thus, in the case that $Q = 1$, or

$$Q_{mw} = \frac{\sqrt{1-k_{eff}^2}}{k_{eff}} \qquad (3.66)$$

virtually exact tuning takes place in the band between frequencies that correspond to ΔC_{max} and ΔC_{min} around the resonance frequency. Under the condition that $Q > 1$ (Q_1 in Figure 3.9), the tanexact tuning is achieved at three points as it was mentioned before, and between these points (f_{t1}, f_r, and f_{t2} in Figure 3.9) a detuning takes place, which is characterized by formula (3.62) for $\cos\varphi$. The actual magnitude of detuning depends on how strong the inequality $Q_1 > 1$ is. In the case that $Q < 1$, the resonance frequency is the only frequency of exact tuning unless a different frequency is chosen for this purpose.

The value $Q = 1$ and relation (3.66) between Q_{mw} and k_{eff} can be considered as optimal for ideal (with the smallest deviation from $\cos\varphi = 1$) tuning in a broad frequency band around the resonance frequency. At the same time a sufficiently good tuning can be achieved in a broader frequency band, if to admit deviation of the power factor to a smaller value. Usually $\cos\varphi \approx 0.8$ is considered acceptable.[9] As it follows From Figure 3.9 at $Q > Q_{opt}$ the most deviation from the optimal tuning takes place at frequencies $f_r \mp \Delta f_{0.5}$, at which $\pm \Delta C_{p\,max}$ is achieved. The situation with detuning that is characterized by deviation $\Delta(\omega)$ from value of function $F(f)$ is symmetrical and can be considered for frequency $f_r - \Delta f_{0.5}$ for example. At this frequency

$$F(f_r - \Delta f_{0.5}) + \Delta(\omega_r - \Delta \omega_{0.5}) = \Delta C_{p\,max}. \qquad (3.67)$$

The deviation $\Delta(\omega_r - \Delta \omega_{0.5})$, at which $\cos\varphi = 0.8$, will be found from Eq. (3.62) as

$$\Delta(\omega_r - \Delta \omega_{0.5}) = 0.75 \frac{G(\omega_r - \Delta \omega_{0.5})}{\omega_r - \Delta \omega_{0.5}} \approx 0.37 G_{mw} \frac{1}{\omega_r(1 - \Delta f_{0.5}/f_r)} \qquad (3.68)$$

or, given that $G_{mw} = n^2/r_{mw} = \omega_r Q_{mw} C_{eqv}^E n^2$,

$$\Delta(\omega_r - \Delta \omega_{0.5}) = 0.37 Q_{mw} C_{eqv}^E n^2 \frac{1}{1 - \Delta f_{0.5}/f_r}. \qquad (3.69)$$

According to formula (3.27)

$$\Delta C_{p\max} = \frac{1}{2} Q_{mw}^2 C_{eqv}^E n^2. \qquad (3.70)$$

After substituting into Eq. (3.67) expressions (3.70), (3.69), and (3.60) for $F(f_r - \Delta f_{0.5})$, and after simple manipulations, in process of which relations

$$\frac{\Delta f_{0.5}}{f_r} = \frac{1}{2Q_{mw}} \quad \text{and} \quad \frac{C_{eqv}^E n^2}{C_{el}^S} = \frac{k_{eff}^2}{1 - k_{eff}^2} \qquad (3.71)$$

are taken into consideration, we will obtain equation for determining the quality factor Q_{mw}

$$Q_{mw} = 0.87 + \sqrt{0.88 + 2\frac{1 - k_{eff}^2}{k_{eff}^2}}. \qquad (3.72)$$

Thus, for example, for the pulsating sphere with $k_{eff} = k_p = 0.58$ $Q_{mw\,opt} = 1.4$ and Q_{mw}, at which within points of intersection of functions $C_p(\omega)$ and $F(\omega)$ the power factor is $\cos\varphi \geq 0.8$, is $Q_{mw} = 3.1$.

Consider now peculiarity of tuning the projector operating at frequencies well below the resonance frequency (in the range B shown in Figure 3.7). At frequencies far below the resonance frequency, i.e., at $(\omega/\omega_r) = \Omega_B \ll 1$, formulas (3.19) and (3.20) simplify to

$$R_p = \frac{1 + (\omega C_m R_m)^2}{(\omega C_m)^2 R_m}, \qquad (3.73)$$

$$\Delta C_p = C_m \cdot \frac{1}{1 + (\omega C_m R_m)^2}. \qquad (3.74)$$

If we assume that the mechanical quality factor measured in air, Q_{ma}, does not change with frequency (this is a reasonable assumption for piezoelectric element), then

$$\frac{1}{\omega C_m^E r_{mL}} = Q_{ma}. \qquad (3.75)$$

If to take into consideration that r_{ac} is assumed to be constant, as far as operation in a large array is concerned, then after following manipulations

$$\omega C_m R_m = \omega C_m^E (r_{mL} + r_{ac}) = \frac{1}{Q_{ma}} + \Omega_B \omega_r C_m^E r_{ac} = \frac{1}{Q_{ma}} + \Omega_B \frac{1}{Q_{mw}}. \qquad (3.76)$$

it becomes obvious that $(\omega C_m R_m)^2 \ll 1$ (note that this inequality does not change, if to assume that r_{ac} drops with frequency), and equations (3.73) and (3.74) can be reduced to

3.1. Operation in Transmit Mode

$$R_p \doteq \frac{1}{(\omega C_m)^2 R_m}, \quad \Delta C_p \doteq C_m = C_m^E n^2. \tag{3.77}$$

Respectively,

$$C_p \doteq C_e^s + C_m^E n^2 = C_e^s / (1 - k_{eff}^2), \tag{3.78}$$

$$G_p = \frac{1}{R_{eL}} + R_m (\omega C_m)^2 = \frac{1}{R_{eL}}\left[1 + \frac{R_{eL}}{R_m}\left(\frac{1}{Q_{ma}} + \Omega_B \frac{1}{Q_{mw}}\right)^2\right]. \tag{3.79}$$

Recalling that $R_m = (r_{mL} + r_{ac})/n^2$, the terms involved in relation (3.79) can be represented as follows

$$R_{eL} = \frac{Q_e}{\omega C_e^S} = \frac{1}{\Omega_B} \cdot \frac{Q_e}{\omega_r C_e^S}, \quad r_{mL} = \frac{1}{\omega C_m^E Q_{ma}} = \frac{1}{\Omega_B} \cdot \frac{1}{\omega_r C_m^E Q_{ma}}, \quad r_{ac} \approx \frac{1}{\omega_r C_m^E Q_{mw}}. \tag{3.80}$$

After some manipulations it will be obtain that

$$\frac{R_{eL}}{R_m} = \frac{R_{eL} n^2}{r_{mL} + r_{ac}} = Q_e \cdot \frac{k_{eff}^2}{1 - k_{eff}^2}\left(\frac{1}{Q_{ma}} + \Omega_B \frac{1}{Q_{mw}}\right)^{-1}, \tag{3.81}$$

and finally, we arrive at the following expression for G_p

$$G_p = \frac{\omega C_e^S}{Q_e}\left[1 + Q_e \frac{k_{eff}^2}{1 - k_{eff}^2}\left(\frac{1}{Q_{ma}} + \frac{\Omega_B}{Q_{mw}}\right)\right]. \tag{3.82}$$

Now we can determine Q_{ew} and $\cos\varphi$ for the not tuned input impedance as

$$Q_{ew} = \frac{\omega C_p}{G_p} = \frac{Q_e}{1 + \frac{k_{eff}^2}{1 - k_{eff}^2} \cdot \frac{Q_e}{Q_{ma}}\left(1 + \Omega_B \frac{Q_{ma}}{Q_{mw}}\right)} \cdot \frac{1}{1 - k_{eff}^2} =$$

$$= \frac{Q_e}{1 - k_{eff}^2 + k_{eff}^2 \frac{Q_e}{Q_{ma}}\left(1 + \Omega_B \frac{Q_{ma}}{Q_{mw}}\right)} \tag{3.83}$$

and

$$\cos\varphi = \frac{1}{\sqrt{1 + Q_{ew}^2}}. \tag{3.84}$$

Consider now electrical circuit of parallel tuning of a projector shown in Figure 3.6. The inductance L in this circuit should be

$$L = \frac{1}{\omega_t^2 C_p}, \qquad (3.85)$$

where ω_t is the frequency of exact tuning and $R_L = Q_L \omega L$ is the equivalent resistance of losses in the inductor, having quality factor Q_L. The conductivity, G_{tc}, of the tuning circuit at the frequency of exact tuning will be found as

$$G_{tc} = G_p + \frac{1}{R_L} = G_p + \frac{1}{\omega_t L Q_L}, \qquad (3.86)$$

After substituting $G_p = \omega C_p / Q_{ew}$ in this relation (see formula (3.83)) and L from formula (3.85) we obtain

$$G_{tc} = \omega_t C_p \left(\frac{Q_{ew} + Q_L}{Q_{ew} Q_L} \right). \qquad (3.87)$$

We will introduce the electric quality of the tuning circuit, Q_{tc}, as

$$Q_{tc} = \frac{\omega_t C_p}{G_{tc}} = \frac{Q_{ew} Q_L}{Q_{ew} + Q_L}, \qquad (3.88)$$

and the efficiency of the tuning circuit, η_{tc}, as

$$\eta_{tc} = \frac{\dot{W}_{el_{tr}}}{\dot{W}_{el_{tc}}}, \qquad (3.89)$$

where $\dot{W}_{el_{tc}} = V^2 G_{tc}$ is the total active power supplied to the tuning circuit and $\dot{W}_{el_{tr}} = V^2 G_p = V^2 \omega_t C_p / Q_{ew}$ is the active power consumed by the projector. Thus,

$$\eta_{tc} = \frac{Q_L}{Q_L + Q_{ew}} = \frac{1}{1 + Q_{ew} / Q_L}. \qquad (3.90)$$

We will assume for estimations that $Q_L = 100$, although, as it was noted before, this value may be too optimistic for a practical inductor design, given that the magnitude of operating voltage across the inductor may be of the order of kilovolts.

Now consider the power factor, $\cos\varphi_{tc}$, of the input impedance of the tuned projector vs. frequency deviation from the frequency of exact tuning. By definition

$$\cos\varphi_{tc} = \frac{I_G}{|I|} = \frac{G_{ct}}{\sqrt{G_{tc}^2 + |Y|^2}}, \qquad (3.91)$$

where $|Y| = \omega C_p (1 - f_t^2 / f^2)$. If to represent the operating frequency as $f = f_t(1 \pm \Delta f / f_t)$ then this relation may be reduced to

$$\cos\varphi_{tc} = \frac{1}{\sqrt{1 + (2\Delta f / f_t)^2 Q_{tc}^2}}, \qquad (3.92)$$

where from follows that the power factor of a tuned projector drops down to 0.7 at deviation from the tuned frequency

$$\frac{\Delta f_{0.7}}{f_t} = \frac{1}{2 Q_{tc}}. \qquad (3.93)$$

If, for example, $Q_L = 100$ and $Q_{ew} = 60$ (the data can be considered as representative of a projector operating at low frequencies) $\cos\varphi_{tc} = 0.7$ at the deviation of frequency $\Delta f_{0.7} \approx 0.01 f_t$ ($\cos\varphi_{tc} = 0.3$ at $\Delta f_{0.3} \approx 0.04 f_t$).

Thus, the frequency band of acceptable tuning in terms of the value of the power factor of the transmit channel is very narrow. Note that usually $Q_L \gg Q_{ew}(0)$ and in a narrow band around the resonance frequency the term Q_{ew}/Q_L in formulas (3.88) and (3.90) can be neglected. But in the case of a broad band operation, firstly, $Q_{ew}(\Omega)$ vs. Ω may increase rapidly and, secondly, given that voltage across the tuning inductor can be fairly large, in a broad band operation it may become not affordable to get a high Q_L.

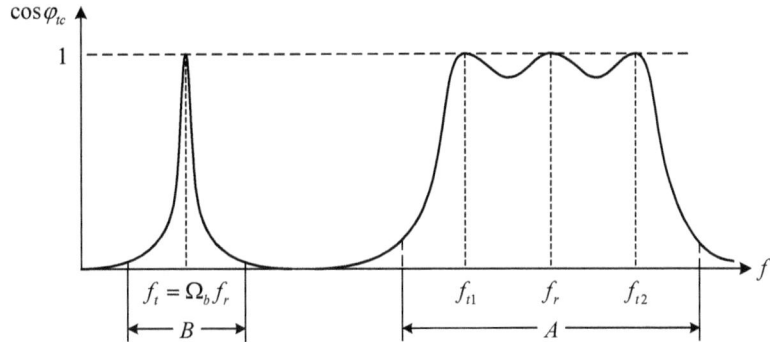

Figure 3.10: Qualitative illustration of difference in results of tuning in the frequency ranges A and B.

Qualitative illustration of difference in results of tuning in the frequency range around the resonance frequency and far below this range is presented in Figure 3.10.

Estimations can be made of the optimal for tuning values of the quality factors Q_{mw} using relation (3.66). Suppose that the projectors are made of PZT-4 ceramics, for which the coupling coefficients are[2]: $k_p = 0.58$, $k_{31} = 0.33$, $k_{33} = 0.70$. These coupling coefficients are typical for the spherical transducer and for the cylindrical transducer solid and segmented, respectively. For the bar transducer the effective coupling coefficients must be used. For the solid bar they are $k_{eff1} = 0.3$ and $k_{eff3} = 0.57$ for the side electroded and for end electroded piezoelements, respectively, according to relations (2.94). The corresponding optimal values of the quality factors are: 1.4 for sphere, 2.9 and 1.0 for the solid and segmented rings, 2.8 and 1.4 for the solid side and end electroded bars. Comparison with data presented in Table 3.1 on the quality factors Q_{mw} leads to the conclusions that in terms of optimal tuning all the transducers are acoustically under loaded ($Q_{we} < Q_{opt}$). The closest to optimal is loading of the solid cylindrical transducer, and the most under loaded is the bar transducer. For spheres and cylinders a better matching with acoustic load can be achieved by reducing their thickness relative to radius. Conditions of loading for the bars cannot be changed without using additional matching elements in their design. Employing the Tonpilz design is the most efficient way of matching bar transducer with acoustic load. This issue will be discussed in Section 10.4.1.

Estimations made in this section are not rigorous, because values of the resistances of the mechanical and electrical losses that play a crucial role are not known to great accuracy. Thus, they show the qualitative tendencies.

The tuning situation can be easily modeled numerically so far as parameters of the input impedance are known, and appropriate value of inductance thus can be determined. Moreover, more complicated tuning circuits can be used for improving quality of the tuning, as it is shown for example in Ref. 9. The transducer related designing problem is in achieving a proper value of the quality factor Q_{mw}.

It must be noted that the tuning issues are not the only considerations that determine the frequency band of a projector operation, in which the requirements for the transmit channel must be met ("bandwidth" of the projector). The final conclusions regarding the necessary band of tuning must be made after all the issues related to determining the transducer bandwidth will be discussed.

3.1.2 Effectiveness Factor of the Transmit Channel and Efficiency of a Projector

The goal of a transmit channel as it was formulated in Section 3.1 is in providing a certain acoustic intensity on the acoustical axis, $I(0)$, and a required spatial distribution of sound pressure described by the directional factor, $H(\theta,\phi)$. Assuming that requirements for the directional factor are satisfied, the value of $I(0)$ referenced to 1 meter from the acoustic center fully characterizes useful acoustic output of the transmit channel.

Thus, the quality of converting the input electric energy, $\dot{W}_{e.ch}$, into a useful acoustical output performed by a transmit channel may be characterized by the coefficient

$$Ef_{ch} = \frac{I(0)}{\dot{W}_{e.ch}}, \qquad (3.94)$$

which we will define as the *effectiveness factor* of the transmit channel. It is convenient for the further analysis to break the effectiveness factor for the channel into two factors related to the driving amplifier and to the projector,

$$Ef_{ch} = \frac{\dot{W}_{e.tr}}{\dot{W}_{e.ch}} \cdot \frac{I(0)}{\dot{W}_{e.tr}}, \qquad (3.95)$$

where the first factor

$$\eta_{da} = \frac{\dot{W}_{e.tr}}{\dot{W}_{e.ch}} \qquad (3.96)$$

is the efficiency of the driver amplifier loaded by the projector's input impedance. It can be assumed that the input impedance is tuned, as considered in the previous section, and matched with the internal impedance of the amplifier. The second factor

$$Ef_{tr} = \frac{I(0)}{\dot{W}_{e.tr}} \qquad (3.97)$$

is the effectiveness of the projector. The more common characterization of the energy transmission by a projector is the electroacoustic efficiency, η_{ea}, which is defined as

$$\eta_{ea} = \frac{\dot{W}_{ac}}{\dot{W}_{e.tr}}. \qquad (3.98)$$

where $\dot{W}_{ac}$ is the total acoustic power radiated. Assuming that directivity D of the projector is determined as[7]

$$D = \frac{4\pi}{\int_{4\pi} |H(\theta,\phi)|^2 d\Omega} = 4\pi \frac{I(0)}{\dot{W}_{ac}}, \qquad (3.99)$$

(where Ω is the solid angle), relation between the effectiveness and electroacoustic efficiency of the projector may be expressed as

$$\eta_{ea} = \frac{4\pi}{D \cdot Ef_{tr}}. \qquad (3.100)$$

The effectiveness factor has several advantages over the efficiency η_{ea}. Firstly, the effectiveness is accounted for the "useful" part of the total acoustic power that is radiated in the direction of the acoustic axis only, whereas the total acoustic power includes radiation in unwanted directions, which is a waste of energy from the viewpoint of a user of the transmit channel. Secondly, it is easy to measure the sound pressure and therefore the intensity in a certain direction, whereas it is hard to accurately measure the total acoustical power radiated.

One of the methods for determining the acoustic power (the direct method) presupposes measuring the intensity on acoustical axis and evaluating the directivity, which in its turn involves measuring the directional factors of the projector in several planes and subsequent calculations. The whole procedure is time consuming and subject to considerable errors. Another method (the impedance method) is based on analysis of balance between the components of the equivalent electromechanical circuit of a projector, which are responsible for the electrical and mechanical losses in the projector, and the radiation resistance that represents the acoustic load. A fairly simple experimental application of this method was discussed in the preceding article. Though more practical in terms of realization, this method has its shortcomings because the impedances involved may not be accurately represented. In particular, when measuring the radiation resistance, it is hard to separate the part, which represents a "truly" acoustic radiation from the acoustic losses due to radiation into "wet" elements of transducer design (such as the baffles, for example). Besides, the "useful" radiation (in direction of the acoustical axis) constitutes only a part of the total acoustic radiation. As the result, it is very likely that thus measured efficiency might be falsely increased. Introducing the effectiveness makes it possible to avoid all the complications and inaccuracies involved in determining the efficiency, η_{ea}, and the effectiveness factor seems to be a preferable characterization of the projector power handling capacity.

On the other hand, the efficiency η_{ea} is a good characterization of a transducer as an electromechanical system, because it accounts for energy losses within the transducer mechanical system that may become a limiting factor for power radiated by causing the transducer overheating, and its analysis may help to improve the transducer design

3.1.2.1 Efficiency of a Projector

The efficiency of a projector formally may be determined from the equivalent circuit in Figure 3.1. Expression (3.98) for the electroacoustic efficiency can be represented as

$$\eta_{ea} = \frac{\dot{W}_{ac}}{\dot{W}_{e.tr}} = \frac{\dot{W}_{ac}}{\dot{W}_{m}^{E}} \cdot \frac{\dot{W}_{m}^{E}}{\dot{W}_{e.tr}} = \eta_{ma} \eta_{em}, \tag{3.101}$$

where

$$\eta_{em} = \frac{\dot{W}_{m}^{E}}{\dot{W}_{e.tr}} \tag{3.102}$$

is the electromechanical efficiency of a projector, and

$$\eta_{ma} = \frac{\dot{W}_{ac}}{\dot{W}_{mech}} \tag{3.103}$$

is the mechanoacoustic efficiency of the projector. The expressions for the active energies involved in the definitions (3.102) and (3.103) follow directly from the equivalent circuit in Figure 3.1, namely,

$$\dot{W}_{m}^{E} = \frac{|V_n|^2}{|Z_m^E|^2}(r_{ac} + r_{mL}), \tag{3.104}$$

$$\dot{W}_{e.tr} = |V|^2 \left[\frac{1}{R_{eL}} + \frac{n^2}{|Z_m^E|^2}(r_{ac} + r_{mL}) \right]. \tag{3.105}$$

After substituting the expressions for energies into the formulas (3.102) and (3.103) we arrive at the following representations for the efficiencies

$$\eta_{em} = \left\{ 1 + \frac{r_{ac} + r_{mL}}{n^2 R_{eL}} \left[1 + Q_{mw}^2 \left(\frac{f}{f_{mw}} - \frac{f_{mw}}{f} \right)^2 \right] \right\}^{-1}, \qquad (3.106)$$

and

$$\eta_{ma} = \frac{r_{ac}}{r_{mL} + r_{ac}}. \qquad (3.107)$$

It follows from expression (3.106) that η_{em} strongly depends on the frequency deviation from the resonance frequency of the projector. As for η_{ma}, it can be considered practically independent of frequency so far as r_{mL} and r_{ac} change only slightly in the narrow frequency band around the resonance frequency, and far enough of resonance frequency the very rapid changes of η_{em} are dominating anyway. Thus, we can assume that

$$\eta_{ma} = \eta_{ma\,r}, \qquad (3.108)$$

where $\eta_{ma\,r}$ is the mechanoacoustic efficiency at the resonance frequency of the projector. At the resonance frequency expression (3.101) for electroacoustic efficiency simplifies to

$$\eta_{ea\,r} = \eta_{ma\,r} \eta_{em\,r} = \frac{r_{ac}}{r_{ac} + r_{mL}} \cdot \frac{1}{1 + (r_{ac} + r_{mL})/(n^2 R_{eL})}. \qquad (3.109)$$

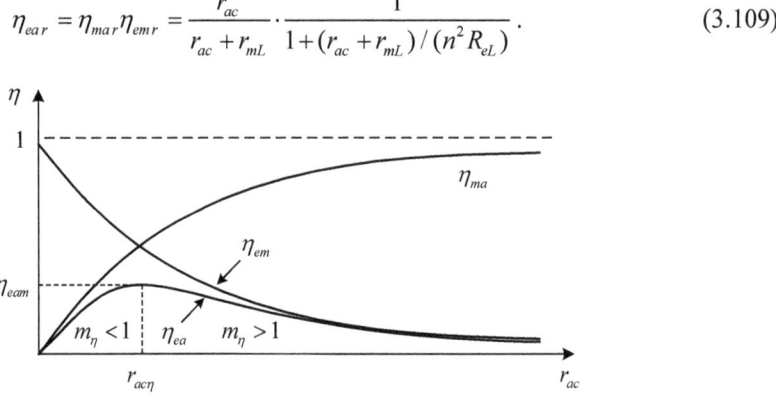

Figure 3.11: Efficiencies of a projector vs. acoustic load.

All the efficiencies of a projector depend on the acoustic load. Their dependencies on the radiation resistance are illustrated qualitatively in Figure 3.11 (subscript "r" is omitted). While η_{ma} increases with rise of the radiation resistance, η_{em} decreases, and η_{ea} has a maximum, $\eta_{ea\,m}$, at some value of r_{ac}, which we denote as $r_{ac\eta}$. This value of the radiation resistance can be found from the condition

3.1. Operation in Transmit Mode

$$\left.\frac{\partial \eta_{ea}}{\partial r_{ac}}\right|_{r_{ac}=r_{ac\eta}} = 0, \qquad (3.110)$$

where η_{ea} is given by Eq. (3.109). After calculations, we arrive at

$$r_{ac\eta} = r_{mL}\sqrt{1 + n^2 R_{eL}/r_{mL}}. \qquad (3.111)$$

If the parameters of a particular projector involved in formula (3.111) are known, then the optimal $r_{ac\eta}$ and the maximum available efficiency of the projector, which will be obtained by substituting $r_{ac\eta}$ in formula (3.109), can be determined. Although in practice the freedom to vary r_{ac} is restricted, the conclusion can be made on what kind of losses are critical under a particular acoustic loading in terms of the overall efficiency improvement. The ratio of an operating value of r_{ac} to its optimal value $r_{ac\eta}$ is denoted as

$$m_\eta = \frac{r_{ac}}{r_{ac\eta}} \qquad (3.112)$$

The coefficient m_η characterizes mismatch of the acoustic load in terms of maximizing the electroacoustic efficiency. From Figure 3.11 follows that in the case that $m_\eta > 1$ (the projector is overloaded) the electrical losses are predominant, and in the case that $m_\eta < 1$ (the projector is under loaded) the mechanical losses prevail.

The resistances R_{eL} and r_{mL} of a projector at the resonance frequency can be represented as

$$R_{eL} = \frac{Q_{ea}}{\omega_r C_e^s} \text{ and } r_{mL} = \frac{1}{\omega_r C_m^E Q_{ma}}. \qquad (3.113)$$

Here Q_{ea} and Q_{ma} are the quality factors measured in air, because for the projector operating in water radiation resistance will be added. Upon substituting these expressions into formula (3.111) and noting that $n^2 C_m^E / C_e^s = k_{eff}^2 / (1 - k_{eff}^2)$ we obtain

$$r_{ac\eta} = r_{mL}\sqrt{1 + [k_{eff}^2/(1-k_{eff}^2)]Q_{ma}Q_{ea}} = r_{mL} A_\eta. \qquad (3.114)$$

In this formula the dimensionless factor

$$A_\eta = \sqrt{1 + [k_{eff}^2/(1-k_{eff}^2)]Q_{ma}Q_{ea}} \qquad (3.115)$$

is introduced that is a function of the internal losses in the projector and of its effective coupling coefficient. Now the maximum efficiency of the projector at optimal acoustic load, η_{eam}, can be estimated as

$$\eta_{eam} = (A_\eta - 1)/(A_\eta + 1), \qquad (3.116)$$

following formula (3.109). The rate of change of electroacoustic efficiency relative to its maximum value at the optimal acoustic load vs. mismatch between a real load and its optimal value, $\eta_{ea}(m_\eta)/\eta_{ea}(1)$, depends on value of coefficient A_η.

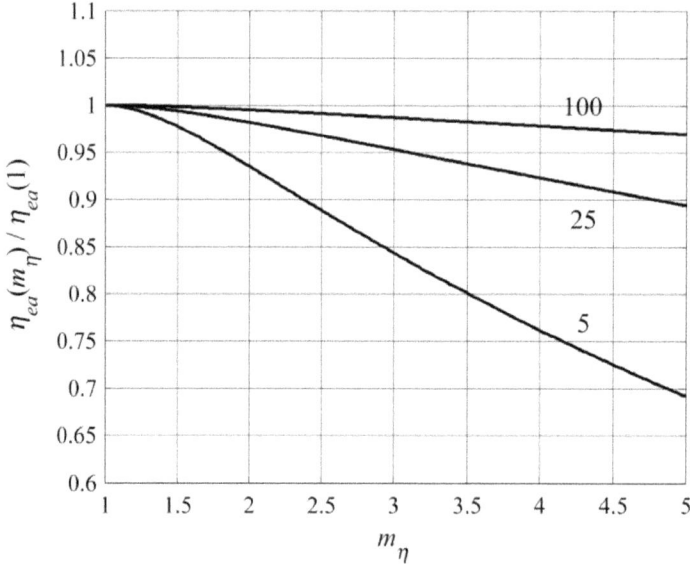

Figure 3.12: Plots of the relative change of the electroacoustic efficiency vs. mismatch between the real and optimal loads. The value of A_η is shown in the plot.

This is illustrated with plots in Figure 3.12. It must be noted that the measurements of the quality factors Q_{ma} and Q_{ea} of a projector in air must be made with caution. One has to make sure that mechanical stress in the projector does not exceed the permissible value, T_p, as Q_{ma} can be much greater than Q_{mw} under real acoustic load in water. Therefore, the full operating electric field cannot be applied when measuring in air if a projector for radiation a great power is concerned. The results obtained in this case by measuring at low electrical field cannot be extrapolated to their maximum permissible values, due to the nonlinear nature of the mechanism of energy dissipation.

3.1. Operation in Transmit Mode

It is interesting to make at least a rough analytical comparison between transducers of different kinds that employ different piezoceramics in terms of their potential maximum efficiencies. With this goal we will consider situation that is idealized, as follows.

The quality factors Q_{ma} and Q_{ea} of a real finished transducer account for additional mechanical and electrical losses that can be considered as parasitic (stray) losses, if compared with the internal losses in a bare piezoelement, which are inherent in a piezoelectric material. The latter are characterized by the quality factors $Q_e = Q_{es}(1-k_{eff}^2)$ and Q_m. We will assume that the stray losses may be neglected. To some extent this can be achieved by a rational transducer designing. The internal dielectric and mechanical losses may be lesser quantitatively than the stray losses, but they directly contribute to a heat release inside of the piezoelement and may become a limiting factor for the maximum power radiated because of an overheating of the piezoelements. Moreover, Q_{es} and Q_m behave in a nonlinear way and at large electrical field and mechanical stress the percentage of energy dissipated may increase significantly. Though the estimations made under such simplifying assumption should result in exaggerating the efficiency, they are useful for illustrating the existing tendencies.

Thus, formula (3.107) reduces to

$$A_\eta = \sqrt{1+k_{eff}^2 Q_m Q_{es}} = \sqrt{1+k_{eff}^2 Q_m / \tan\delta}\,. \tag{3.117}$$

The values of quality factors Q_m and $Q_{es} = 1/\tan\delta$ of the piezoceramic materials at high electrical field and mechanical stress can be found in catalogues issued by the manufacturers of piezoelements. Data regarding values of Q_m and $\tan\delta$ related to low and high electric fields and mechanical stress are presented in Table 3.2 for the piezoceramic compositions PZT-4, PZT-8 and PZT-5A (though the "soft" PZT-5A composition is not supposed to be used for the powerful projectors, this material is included for comparison). Results of calculations made with examples of transducers considered in Chapter 2 that illustrate differences of values of the projector's efficiencies are presented in Table 3.3. The solid rings at the transverse piezoeffect (i. e., with k_{31} coupling coefficient) are taken into calculation only. Therefore, the results are representative for relatively high frequency transducers (for rings at $f > 10$ kHz). The relatively low frequency transducer designs usually employ segmented mechanical systems, in which case some additional mechanical losses due to cementing the piezoelements are involved.

Table 3.2: Properties of the piezoceramic compositions. Presented are the averaged data taken from catalogues of the EDO Corporation, Channel Industries Inc., Morgan and Matroc and from Ref. [2]

		PZT-4	PZT-8	PZT-5A
$\tan\delta = \dfrac{1}{Q_{es}}$	Low field	0.004	0.003	0.02
	$E = 2$ kV/cm (5 V/mil*)	0.02	0.005	0.16
Q_m	Low stress	500	1000	75
	$T = 20$ MPa (3000 psi**)	100	600	30

*1 V/mil = 0.4 kV/cm   **1 psi = 6895 Pa

Table 3.3: Results of calculating the efficiencies of projectors.

	Material	Low Field			$E = 2$ kV/cm		
		A_η	m_η	η_{eam}	A_η	m_η	η_{eam}
$a/t = 5$, $Q_{mw} = 3.6$	PZT-4	117	1.19	0.98	24.5	1.25	0.92
	PZT-8	150	1.85	0.99	95.0	1.46	0.98
	PZT-5A	20.8	1.00	0.91	4.3	1.62	0.62
$a/t = 5$, $Q_{mw} = 7.2$	PZT-4	117	0.60	0.98	24.5	0.62	0.92
	PZT-8	150	0.93	0.99	95.0	0.73	0.98
	PZT-5A	20.8	0.50	0.89	4.3	0.81	0.62

Data presented in Table 3.3 show the potential maximum values of the efficiencies, η_{eam}, and the mismatch coefficients, m_η. They change depending on the driving conditions. In order to estimate how the values of efficiencies change at acoustic loads that correspond to the mismatch coefficients from the table, the plots in Figure 3.12 may be used. In the idealized case, in which only losses in the ceramics are taken into consideration, coefficients A_η are large enough for effects of the mismatch being negligible with exception of transducers made of PZT-5A. For these transducers under hard drive the additional drop of efficiency due to the mismatch will be about 5%. The effect of the mismatch on efficiency of transducers with real internal losses taken into account may be much more significant.

3.1.2.2 Efficiency of a Projector over a Frequency Band

After substituting expressions (3.106)-(3.108) into formula (3.101) we will arrive at the relation $\eta_{ea}(\Omega) = \eta_{ea\,r} B(\Omega)$, where Ω characterizes deviation from the resonance frequency (see definition (3.15)), and

$$B(\Omega) = \frac{1}{\eta_{em\,r} + (1-\eta_{em\,r})(1+\Omega^2 Q_{mw}^2)}. \qquad (3.118)$$

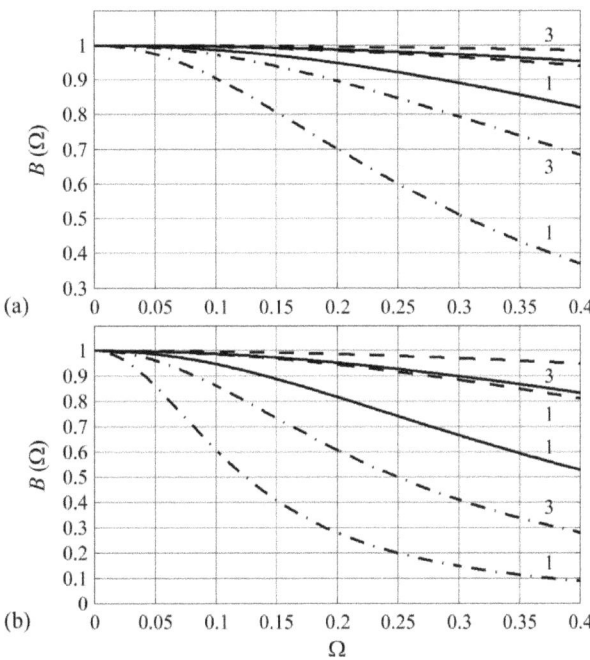

Figure 3.13: Plot of function $B(\Omega) = \eta_{em}(\Omega)/\eta_{em\,r}$ vs. frequency band for PZT-4 (solid lines), PZT-8 (dashed lines) and PZT-5A (dot-dashed lines): (a) cylindrical transducer ($Q_{mw} = 7.2$), (b) bar transducer ($Q_{mw} = 25$). Labels 1 are for the case $k_{eff} = k_{31}$ and labels 3 are for the case $k_{eff} = k_{33}$.

Thus, analysis of the electroacoustic efficiency in a broad frequency band may be reduced to the analysis of the function $B(\Omega)$. This function depends essentially on the value of $\eta_{em\,r}$ and, in the case that $\eta_{em\,r}$ is close to unity, $B(\Omega)$ becomes nearly independent of the frequency. Note that under the condition that maximum power has to be radiated over the entire operating band efficiency $\eta_{em\,r}$ must be calculated with value of $\tan\delta$ corresponding to the permissible electric field $E_p = 2 \cdot 10^5$ V/m on the border of the band.

Consider a numerical estimation of the function $B(\Omega)$ vs. frequency deviation using the examples of the transducers, for which values of Q_{mw} under a "natural" acoustic load are: for the cylindrical transducer $Q_{mw} = 7.2$ at $a/t = 5$, and for the bar transducer $Q_{mw} = 25$. Behavior of the function $B(\Omega)$ is illustrated in Figure 3.13.

The dot-dashed lines in Figure 3.13 show how the efficiency of the transducers made of "hard" ceramic materials would change, if they were made of "soft" PZT-5A ceramic composition.

It must be recognized judging by the results presented that decrease of efficiency of transducer itself is not a limiting factor for a broadband operation, if a proper PZT material is used (so far as the heat release considerations are not concerned). The effectiveness of a transmit channel may be affected more seriously by efficiency of the driving amplifier together with tuning and matching elements in operation on the transducer impedance that becomes highly reactive and frequency dependent.

3.1.3 Maximum Acoustic Power Radiated by a Transducer and Its Limitations

The ability to deliver the maximum possible acoustic power is the most important characterization of a projector. This is obvious in the case that it is needed by a requirement. But even if this is not required, and the projector is intended for radiating a moderate power, comparing the operating power with the maximum possible shows, how much of the reserves available can be used for rationalizing the projector design for a particular application. The maximum acoustic power is limited by the mechanical and electrical strength of the transducer, as well as by possible overheating caused by energy dissipation within the transducer. The latter will not be considered on a general level, because the thermal conditions depend very much on the operating regime of a transducer and on the detail of a particular transducer design.

The acoustic power radiated, being expressed from the equivalent circuit of Figure 3.1, may be represented after using relation (3.18) for the impedance of mechanical system as

$$\dot{W}_{ac} = \left|\frac{Vn}{Z_m^E}\right|^2 r_{ac} = \frac{(Vn)^2 r_{ac}}{(r_{mL}+r_{ac})^2(1+Q_{mw}^2\Omega^2)} . \tag{3.119}$$

3.1. Operation in Transmit Mode

The maximum acoustic power may be radiated at the resonance frequency of a transducer, and its value is

$$\dot{W}_{acm} = \left(\frac{V_{max} n}{r_{ac} + r_{mL}}\right)^2 r_{ac} = |U_{omax}|^2 r_{ac}. \quad (3.120)$$

As it follows from Eq. (3.120), the acoustic power can be limited by the permissible value of voltage applied, V_{max}, or by the permissible value of the mechanical stress generated in the transducer body, which is proportional to velocity U_{omax}.

Table 3.4: Allowable compressive and rated tensile strss for piezooceramics.

	One-Dimensional Compression MPa (kpsi)		Tensile Dynamic and Static Strength MPa (kpsi)	Hydrostatic Pressure MPa (kpsi)
	∥	⊥		
PZT-4	84 (12)	56 (8)	24 (3.5)	350 (50)
PZT-8	84 (12)	56 (8)	35 (5.0)	350 (50)
PZT-5A	21 (3)	14 (2)	28 (4.0)	140 (20)

The problem of radiating great power is a complex multidisciplinary problem. Two groups of factors that limit a reliable power radiation can be considered: (1) physical and technological and (2) design related. Physical and technological factors, e.g., the electrical and mechanical strength of piezoelectric and auxiliary materials and piezoelements, the transducers fabrication technique, reliability problems including the fatigue strength of the transducers. Factors related to the rational designing the projectors as electromechanical devices, namely, the effective electromechanical coupling, optimal acoustic loading in terms of reducing operating electric field and/or mechanical stress, whichever is limiting, and possible increase of efficiency of the projector. In this treatment we will consider the latter group of factors assuming that all the physical and technological factors are reduced to the representative values of the permissible electric field, E_p, and mechanical stress, T_p, under which an operation of a projector can be considered as linear and long term reliable. The values of E_p and T_p, which are valid for finished transducer designs, are assumed to be correlated with an existing level of the "technological skills" and thus subject to change in accordance with an improvement of these skills. This can change the quantitative estimations made, but the methodical approach to the problem can remain intact. The permissible values of E_p and T_p are not very certain. For example, the data related to the mechanical strength of the ceramics are presented in Table 3.4 that is taken from Ref. 2.

Judging by the data from different sources available in accessible literature, we will use for our estimations the following values: $E_p = 2$ kV/cm (5 V/mil) and $T_p = 20$ MPa (3000 psi). Here T_p is the permissible tensile stress for a solid piezoelement. In the case of the segmented mechanical systems this value is applicable under the condition that they are prestressed by the stress $T = 20$ MPa in the direction of segmenting.

The operating electric field in a transducer piezoelement, which is required for developing the electromechanical force, Vn, may be represented as

$$E = A_E Vn, \qquad (3.121)$$

where coefficient A_E must be determined for each transducer type. Upon substituting the value of Vn from Eq. (3.121) into the left hand part of Eq. (3.120) we arrive at the expression for the maximum available power radiated limited by the electrical strength

$$\dot{W}_{mE} = \frac{E_p^2}{A_E^2} \frac{r_{ac}}{(r_{ac} + r_{mL})^2}. \qquad (3.122)$$

The maximum operating stress in the mechanical system of a transducer is proportional to magnitude of the reference point velocity and this may be represented at resonance frequency as

$$T = A_T U_o, \qquad (3.123)$$

where A_T is coefficient that is determined by a type of mechanical system and its mode of vibration (note that $T \sim \xi_o = U_o/\omega$, thus A_T includes the resonance frequency of the mechanical system). Coefficients A_E and A_T for the transducers considered in Chapter 2 are presented in Table 3.5. Upon substituting the value of U_o determined from Eq. (3.123) into the right hand side of Eq. (3.122) we arrive at the expression for the maximum available power radiated, limited by the mechanical strength of the transducer

$$\dot{W}_{mT} = \frac{T_p^2}{A_T^2} r_{ac}. \qquad (3.124)$$

The maximum power that can be radiated by a projector is equal to the lesser of values $\dot{W}_{mE}$ and $\dot{W}_{mT}$.

3.1.3.1 The Optimal Acoustic Load and the Maximum Power Radiated

Consider the ratio of the maximum stress limited power to the maximum electric field limited power,

$$\frac{\dot{W}_{mT}}{\dot{W}_{mE}} = \left(\frac{A_E}{A_T}\right)^2 \left(\frac{T_p}{E_p}\right)^2 (r_{ac} + r_{mL})^2 = \gamma. \quad (3.125)$$

The first factor in Eq. (3.125), $(A_E / A_T)^2$, is determined by the type and mode of vibration of the transducer mechanical system and by the electromechanical properties of the piezoelectric ceramic material used (it may be called the design factor). The second factor, $(T_p / E_p)^2$, is related to the existing average technological level of fabricating the piezoelements and the finished transducers (it may be called the technological factor). Note that a poor transducer fabrication can reduce this factor dramatically. The third term depends on the acoustic load (the load factor). In the case that in Eq. (3.125) $\gamma > 1$, the maximum power radiated is limited by the electrical strength and the transducer has a reserve of the mechanical strength. In the case that $\gamma < 1$, the power radiated is limited by the mechanical strength, and the transducer has a reserve of the electrical strength. In the case that $\gamma = 1$, the maximum possible acoustical power can be radiated by the transducer. This optimal situation can be achieved by matching the acoustic load to the properties of a particular transducer. It is shown qualitatively in Figure 3.14, how $\dot{W}_{mE}$ and $\dot{W}_{mT}$ change with respect to acoustic load. Coordinates of the intersection point M of the functions $\dot{W}_{mE}$ and $\dot{W}_{mT}$, which correspond to the maximum power, $\dot{W}_{acm}$, and the value of optimal acoustic load, r_{optW}, depends on both the design factor, (A_E / A_T), and the technological factor, (T_P / E_P). Therefore, the optimum can be achieved not only by direct changing the acoustic load that is usually not an easy task, but also by changing the transducer design and thereby by changing the criterion for matching. The value of optimal acoustic load, r_{optW}, can be found from Eq. (3.125) at $\gamma = 1$, namely,

$$r_{optW} = \frac{A_T E_P}{A_E T_P} - r_{mL} \approx \frac{A_T E_P}{A_E T_P}. \quad (3.126)$$

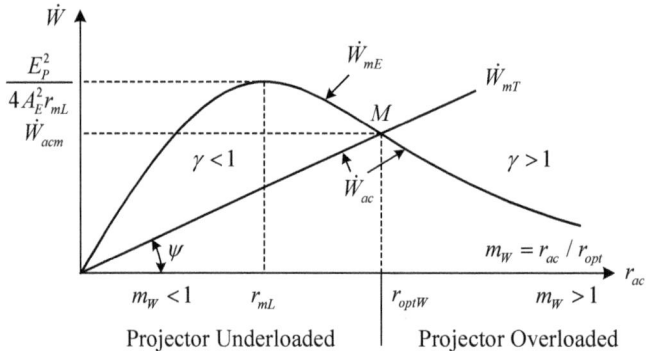

Figure 3.14: The diagram of the maximum acoustic power with respect to acoustic load: $\tan\psi = T_P^2 / A_T^2$.

Resistance of mechanical losses, r_{mL}, can be neglected, because this is equivalent to assuming that at the optimal loading the mechanoacoustic efficiency, η_{ma}, of the transducer is close to unity, as it was illustrated with the typical examples of transducers in Section 3.1.2. As it is shown in Table 3.3, the electroacoustic efficiencies η_{eam} are close to unity (and so do η_{ma} that have greater values). From plots in Figure 3.12 follows that these values do not change significantly in a broad range of mismatch with acoustic load for transducers made of PZT-4 and PZT-8, and the optimal resistances $r_{opt\,\eta}$ and $r_{opt\,w}$ are comparable. The correlation between them is $r_{opt\,w}/r_{opt\,\eta} = m_\eta/m_w$, and from Table 3.3 and Table 3.5 it follows that for the ring transducers $m_\eta/m_w = 1.25/2.7 = 0.46$.

Upon substituting the optimal value of radiation resistance into either equation (3.122) or (3.124) we will obtain for $\dot{W}_{acm}$

$$\dot{W}_{acm} \simeq \frac{E_P T_P}{A_E A_T}. \tag{3.127}$$

The ratio

$$r_{ac}/r_{opt\,W} = m_W \tag{3.128}$$

characterizes a mismatch between the real radiation impedance and its optimal value for a particular transducer. Thus, at $m_W > 1$ in Figure 3.14 the transducer is overloaded, and at $m_W < 1$ it is underloaded. Note that combining Eqs. (3.125), (3.126) and (3.128) results in

$$\gamma = m_w^2. \tag{3.129}$$

3.1. Operation in Transmit Mode

Although the above considerations related to the maximum power radiated by a transducer are quantitatively valid only at its resonance frequency, they are also important in a methodical sense, because they give an estimation of the absolute maximum power obtainable from the transducer and they help to determine the ways for optimizing the transducer design.

Consider cylindrical transducer made of circular rings vibrating in the breathing mode and transducer in the shape of uniform bar vibrating longitudinally in the fundamental mode, as examples for illustrating this statement. In terms of acoustic loading, we will assume that the cylindrical transducer is long compared to the wavelength and the bar transducer operates in a large plane array. The results of estimation of the optimal acoustic load and maximum possible power radiated as well as the maximum power available under the real (natural) acoustic load are presented in Table 3.5.

Table 3.5: Estimations of the optimal acoustic loads and maximum power radiated.

	i	A_E	A_T	r_{opt}/S_Σ, 10^{-2}	r_{ac}/S_Σ [1)]	m_w [2)]	$\dot{W}_{mT}/\dot{W}_{mE}$
Bar	1	s_{ii}^E	$(\rho c_i^E)_c$	$2\dfrac{|d_{3i}|}{s_{ii}^E}(\rho c_i^E)_c$	$0.8(\rho c)_w$	0.24	0.06
	3	$2|d_{3i}|tw$				0.15	0.02
Ring	1	s_{ii}^E	$(\rho c_i^E)_c$	$\dfrac{|d_{3i}|}{s_{ii}^E}(\rho c_i^E)_c\dfrac{t}{a}$	$0.9(\rho c)_w$ at $ka = 2.2$	2.7	7.3
	3	$2\pi|d_{3i}|th$				1.7	2.9

Note: $E_p/T_p = 10^{-2}$ V/mN, $S_\Sigma = 2\pi ah$ for ring, $S_\Sigma = tw$ for bar.

[1)] See Figure 2.4 for ring and Figure 2.10 for bars.
[2)] For ceramics PZT-4; for rings at $t/a = 1/5$.

Several conclusions can be made following the estimations presented in Table 3.5.

The uniform bar transducers are significantly under loaded even operating in a big array, and their power radiated is limited by the mechanical strength. One must pursue an increase of acoustic load for optimizing the transducer design. This can be done at the expense of reduction of the reserve of the electrical strength. Thus, the Tonpilz transducer design can be considered as modification of the uniform bar with the goal of matching acoustic load for obtaining greater acoustic power. The related issues will be considered in Chapter 10.

In contrast, the single cylindrical transducers are overloaded, and their power radiated is limited by the electrical strength. Near to optimal loading of the transducers can be achieved by an increase of the thickness to radius ratio, t/a, and/or by a partial baffling the radiating surface, as is the case for cylindrical transducers used in arrays or as single transducers with baffles for achieving unidirectional radiation in the horizontal plane. This issue will be considered in Chapter 7. The estimations of the maximum available acoustic power radiated are of significant methodical importance. Not only can they help to increase the maximum power radiated by better matching the acoustic load, but they can also be used as guidance for improving designs of transducers intended for radiating a moderate power in a broad frequency band. Introducing the concept of reserves of mechanical and electrical strength may be useful in this respect.

3.1.3.2 Reserves of Strength Coefficients

The operating acoustic power of a projector can be denoted by $\dot{W}_{op}$ and the corresponding electrical field and mechanical stress in the projector as E_{op} and T_{op}. Then the coefficients

$$k_E = \frac{E_P}{E_{op}} = \sqrt{\frac{\dot{W}_{mE}}{\dot{W}_{op}}}, \qquad (3.130)$$

and

$$k_T = \frac{T_P}{T_{op}} = \sqrt{\frac{\dot{W}_{mE}}{\dot{W}_{op}}}, \qquad (3.131)$$

characterize the reserves of the electrical and mechanical strength of the projector regarding the maximum permissible values of the electrical field and mechanical stress. By these definitions for the reserve coefficients, it should be $k_E \geq 1$ and $k_T \geq 1$. The equality corresponds to the case that the operating power reaches the maximum power limited by electrical field or by mechanical stress. If k_E and/or k_T exceeds the unity, it means that the projector has an excessive reserve of strength. In general, the projector design can be regarded as rational in the case that it does not have an excessive reserve of either electrical or mechanical strength, i.e., $k_E = k_T$ under the operating loading conditions. The situation where $k_E = k_T = 1$ corresponds to the maximum power available from projector for a particular application. If a moderate acoustic power is required, then $k_E, k_T > 1$, which means that the projector acquires excessive reserves and its design can be made simpler, cheaper, and even more reliable for expense of a reduction of these reserves (for example, the amount of active material can be reduced, or the prestressing arrangements can be simplified, etc.). Note, that the resulting increase of E_{op} and T_{op} up to their permissible values E_p and T_p should not compromise the reliability of a projector, because the permissible values by their definitions are supposed to insure a long term reliability of a transducer.

3.1.4 Frequency Response of a Projector

In order for a transmit channel to maintain a certain intensity level (3.1) on the acoustic axis in a prescribed frequency band, dependence of the projector voltage sensitivity on the frequency in this frequency band has to be known. Here the projector voltage sensitivity, γ_V, is defined as the ratio of sound pressure generated by the projector on the acoustic axis referenced to 1 m, $P(0)_{1m}$, to voltage applied

$$\gamma_V(\omega) = P(0,\omega)_{1m} / V. \tag{3.132}$$

The sound pressure generated by a projector on the acoustic axis is according to relation (2.28)

$$P(0,\omega) = P_0(U_{\tilde{v}}) k_{dif}(\omega), \tag{3.133}$$

where $P_0(U_{\tilde{v}})$ is the sound pressure generated by the small pulsating sphere having the same volume velocity, as a real transducer, and $k_{dif}(\omega)$ is the diffraction coefficient for the transducer. Following Eq. (2.27)

$$P_0(\omega) = \frac{\rho c}{2\lambda r} U_{\tilde{v}} e^{-j(kr-\pi/2)} = \frac{\rho c S_\Sigma}{2\lambda r} U_o(\omega) e^{-j(kr-\pi/2)}, \tag{3.134}$$

and therefore

$$P(0,\omega)_{1m} = \frac{\rho c S_\Sigma}{2\lambda} U_o(\omega) k_{dif}(\omega) e^{-j(kr-\pi/2)}. \tag{3.135}$$

The magnitude of velocity of the reference center on the projector surface, $U_o(\omega)$, being determined from the equivalent circuit (see Eqs. (1.74) and (1.76)) is

$$U_o(\omega) = \frac{Vn}{Z_m^E} = Vn \frac{1}{j\omega M_{eqv}[1-(f_r/f)^2] + r_{mL} + Z_{ac}}. \tag{3.136}$$

Thus, the general expression for the sound pressure frequency response in a broad frequency band is

$$P(0,\omega)_{1m} = \frac{\rho c S_{\Sigma}}{2\lambda} \frac{Vn}{j\omega M_{eqv}[1-(f_r/f)^2]+r_{mL}+Z_{ac}} k_{dif}(\omega) e^{-j(kr-\pi/2)}. \quad (3.137)$$

It must be understood that in real operation of a transmit channel magnitude of the voltage applied to projector does not remain constant in a frequency band. Its value depends on the properties of a driver amplifier and on conditions of matching its internal impedance with the input impedance of the projector.

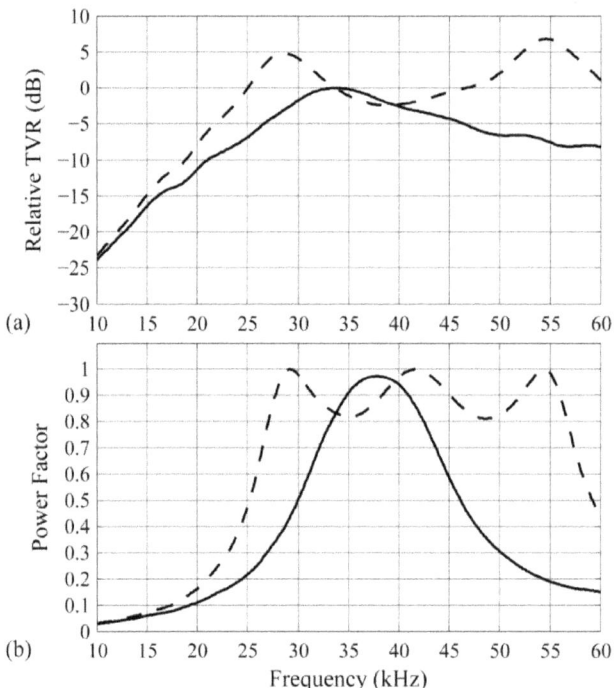

Figure 3.15: Plots of the (a) voltage sensitivity of the spherical transducer (OD = 50.8 mm, t = 2.5 mm, PZT-4) with and without series tuning normalized to its value at resonance frequency and (b) $\cos\varphi_{tr}$ of the transducer. Dashed lines – characteristics of the tuned transducer.

Thus, if the input impedance of the projector is series tuned, the voltage applied to transducer relates to the output voltage of a driver amplifier by relation (3.54) $|V_{tr}|=|V_g|/\cos\varphi_{tr}$, where $\cos\varphi_{tr} = 1/\sqrt{1+Q_{ew}^2}$ (Eq. (3.34)) and $Q_{ew} = \omega|C_p(\omega)|\cdot R_p(\omega)$ (Eq. (3.33)). In the case that internal impedance of the driver amplifier is much smaller than input impedance of the tuned transducer, its output voltage will be kept constant and the frequency response of the tuned transducer will be found as

$$\gamma_{tuned} = \gamma_V / \cos\varphi_{tr}. \tag{3.138}$$

For example, the plots of the voltage sensitivity γ_V, $\cos\varphi_{tr}$ and voltage sensitivity of the tuned spherical transducer are presented in Figure 3.15. Thus, a real frequency response of a tuned transducer can be significantly broader than that of the voltage sensitivity.

As to the voltage sensitivity by formula (3.132) that characterizes property of a projector itself, it must be measured at constant voltage applied to the transducer. Further the voltage sensitivity, γ_V, and its dependence on frequency will be considered. The voltage sensitivity being expressed in the dB scale is called TVR – transmitting voltage response,

$$\text{TVR} = 20 \log P(0) \text{ re } 1 \text{ }\mu\text{Pa/V at 1 m.} \tag{3.139}$$

In the typical operating band of a projector around its resonance frequency the expression for the mechanical impedance can be reduced to the form (3.18). Then Eq. (3.136) becomes

$$U_o(\omega) = \frac{Vn}{(r_{ac} + r_{mL})[1 + j\Omega Q_{mw}]} \approx U_o(\omega_r)\frac{1}{1 + j\Omega Q_{mw}}, \tag{3.140}$$

where

$$U_o(\omega_r) = Vn/(r_{ac\,r} + r_{mL}) = Vn\eta_{ma\,r}/r_{ac\,r}. \tag{3.141}$$

Finally, we arrive at the expression for the frequency response

$$\gamma_V(\omega) = \frac{P(0,\omega)_{1m}}{V} = \frac{\rho c S_\Sigma}{2\lambda}\frac{n}{r_{ac\,r}}\eta_{ma\,r}\frac{1}{1 + j\Omega Q_{mw}}k_{dif}(\omega)e^{-j(kr-\pi/2)}. \tag{3.142}$$

The ratio of its value to value at the resonance frequency is the normalized frequency response of the projector

$$\frac{\gamma_V(f)}{\gamma_V(f_r)} = \frac{f}{f_r}\frac{r_{ac}(f_r)}{r_{ac}(f)}\frac{k_{dif}(f)}{k_{dif}(f_r)}\frac{1}{1 + j\Omega Q_{mw}} \approx \frac{k_{dif}(f)}{k_{dif}(f_r)}\frac{1\pm 0.5\Omega}{1 + j\Omega Q_{mw}}. \tag{3.143}$$

The factor $r_{ac}(f_r)/r_{ac}(f)$ changes insignificantly at deviation $\Omega = \pm 0.4$ from resonance frequency and can be neglected. Thus, from plots in Figures 2.2 and 2.4 follows that for the spherical and cylindrical transducers considered in Chapter 2 $r_{ac}(\Omega)/r_{ac}(0) \approx 1\pm 0.03$). The factor $k_{dif}(f)/k_{dif}(f_r)$ changes slowly with frequency. Its most deviation from resonance value takes place for the spherical transducer and does not exceed ±1.2 at $\Omega = \pm 0.4$.

The normalized frequency response of a projector is an inherent property of a transducer type. It is shown in Figure 3.16 at different values of the quality factor Q_{mw} for the transducers described in Table 3.1.

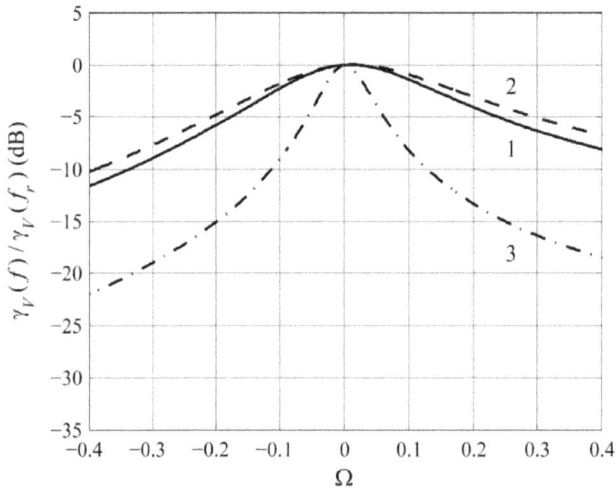

Figure 3.16: Normalized frequency responses of the projectors: (1) cylindrical with a/t = 5, $Q_{mw} = 7.2$, (2) spherical with a/t = 5, $Q_{mw} = 6.0$, and (3) uniform bar with $Q_{mw} = 25$.

The following must be noted regarding the absolute magnitude of the voltage sensitivity of a transducer. Physically effect of generating vibration and hence acoustic radiation is produced by the electric field in the piezoelements comprising the transducer, and the maximum of the sound pressure level (SPL) achievable is limited by the permissible value of the electric field. The sound pressure radiated by a transducer can be changed without affecting the voltage applied. For example, this can be done in the segmented transducer designs by changing the thickness of segments in direction of electric field. (This is one of the ways of matching transducer input impedance that is equivalent to its transformation.) Reducing the thickness of segments produces effect of increase of electric field and of the sound pressure at the same voltage applied to the transducer, i. e., effect of increasing the TVR.

Thus, the TVR cannot be used for comparing radiating abilities of different transducers without providing information about peculiarities of their designs. Much more appropriate for this purpose is TER, the transmit response at unit electric field developed in the comprising piezoelements

3.1.5 Operation of a Projector in a Broad Frequency Band

3.1.5.1 About the Bandwidth of a Projector

The ability of a projector to meet requirements for operating in a broad frequency band is commonly called the transducer Bandwidth. This property of a projector is the most controversial. It cannot be characterized by a single figure of merit, and no single definition can be made that fully describes this concept. Complexity of the bandwidth concept arises from the role of projector, as a member of a transmit channel that has a goal of operating in as wide as necessary frequency band at a certain level of the acoustic power radiated. Not only the projector itself should be able of radiating the required power in the operating frequency band at an acceptable level of efficiency. It has to meet requirements for matching its input impedance with the channel's source of energy that is loaded by the projector as well. Properties of the input impedance and behavior of efficiency of a projector in a frequency band are already considered in the Sections 3.1.1 and 3.1.2. In this section we will be concerned with the projector's ability of radiating a prescribed power in a frequency band around its resonance frequency. In other words, we must be sure that the projector can radiate a required power at each frequency within the specified band around the resonance at maximum deviation of frequency up to $\Omega = \pm 0.4$ (remember that $\Omega = 2\Delta f / f_r$). We denote the acoustic power radiated in this frequency band as $\dot{W}(\Omega)$.

The maximum acoustic power can be radiated by a projector at the resonance frequency, as it was discussed in Section 3.1.3, and the values of acoustic power limited by the electrical and mechanical strengths of the projector are determined by Eqs. (3.122) and (3.124). Here we will denote these values at resonance frequency $\dot{W}_{mE}(0)$ and $\dot{W}_{mT}(0)$. The coefficients of reserves of the electrical and mechanical strength at resonance frequency determined by Eqs. (3.130) and (3.131) will be denoted as $k_E(0)$ and $k_T(0)$, and those in the operating frequency band as $k_E(\Omega)$ and $k_T(\Omega)$.

The acoustic power radiated in operating frequency band is determined by the general expression (3.119)

$$\dot{W}(\Omega) = \frac{[V(\Omega)n]^2 r_{ac}}{(r_{ac} + r_{mL})^2} \cdot \frac{1}{1+\Omega^2 Q_{mw}^2} . \quad (3.144)$$

The values of $r_{ac}(\Omega)$ and $r_{ac}/(r_{ac}+r_{mL})^2$ change slowly compared with strongly frequency dependent second factor of the equation. (This was noted regarding Eq. (3.141)). Taking into consideration that $V \sim E$, the dependence of the acoustic power in the operating frequency band determined by the electric field, $\dot{W}_E$, can be represented as

$$\dot{W}_E(\Omega) = \dot{W}_E(0) \cdot \left[\frac{E(\Omega)}{E(0)}\right]^2 \frac{1}{1+\Omega^2 Q_{mw}^2} , \quad (3.145)$$

and in order to keep the power constant over the frequency band it should be

$$E(\Omega) = E(0)\sqrt{1+\Omega^2 Q_{mw}^2} . \quad (3.146)$$

The expression for the radiated power in the band can be represented as

$$\dot{W}(\Omega) = |U_o(\Omega)|^2 r_{ac}(\Omega) . \quad (3.147)$$

Then

$$\frac{\dot{W}(\Omega)}{\dot{W}(0)} = \frac{|U_o(\Omega)|^2}{|U_o(0)|^2} \cdot \frac{r_{ac}(\Omega)}{r_{acr}} . \quad (3.148)$$

The second factor in this equation is negligible (see the note under Eq. (3.144)). Thua, it can be concluded that for maintaining the power constant in the operating band it should be $U_o(\Omega) \approx U_o(0)$. Given that the mechanical stress is $T \sim \xi_o = U_o/\omega$, the stress should change as

$$T(\Omega) = T(0)\frac{f_r}{f} \approx T(0) \cdot \frac{1}{1\pm(0.5\Omega)} \quad (3.149)$$

(note that $f_r/f = f_r/(f_r \pm \Delta f) \approx 1/(1\pm 0.5\Omega)$). Thus, the stress increases at frequencies below the resonance frequency if the power radiated is kept constant.

To ensure reliable radiation of acoustic power, the requirements $k_E(\Omega) \geq 1$ and $k_T(\Omega) \geq 1$ for values of reserve of strength coefficients must be met. Using expressions (3.146) and (3.149)

3.1. Operation in Transmit Mode

for values of electric field and stress under condition that acoustic power remains constant over the operating band, we obtain the inequalities to be met

$$k_E(\Omega) = \frac{E_p}{E(\Omega)} = \frac{E_p}{E(0)\cdot\sqrt{1+\Omega^2 Q_{mw}^2}} = k_E(0)\cdot\frac{1}{\sqrt{1+\Omega^2 Q_{mw}^2}} \geq 1, \qquad (3.150)$$

$$k_T(\Omega) = \frac{T_p}{T(\Omega)} = \frac{T_p}{T(0)}(1-0.5\Omega) = k_T(0)(1-0.5\Omega) \geq 1. \qquad (3.151)$$

It follows from these inequalities that the projector must possess additional reserves of power available at resonance frequency, which will be denoted $\dot{W}_r(\Omega_m)$, regarding the power $\dot{W}(\Omega)$ required for radiation in the frequency band $\pm\Omega_m$. Namely, the following relations must be fulfilled

$$k_E(0) \geq \sqrt{1+\Omega^2 Q_{mw}^2}, \qquad (3.152)$$

$$k_T(0) \geq \frac{1}{1-0.5\Omega}. \qquad (3.153)$$

The maximum power available in the frequency band $\Omega = \pm\Omega_m$ will be

$$\dot{W}(\Omega) \leq \frac{\dot{W}_r(\Omega_m)}{1+\Omega^2 Q_{mw}^2}. \qquad (3.154)$$

If the maximum power at the resonance frequency is mechanical stress limited, then

$$\dot{W}_r(\Omega_m) = \dot{W}_{mT}(1-\Omega_m). \qquad (3.155)$$

If the maximum power at resonance frequency is electric field limited, then

$$\dot{W}_r(\Omega_m) = \frac{\dot{W}_{mE}}{1+\Omega_m^2 Q_{mw}^2}. \qquad (3.156)$$

We will further consider the case that the maximum power radiated at the resonance frequency is electric field limited, because the stress limited projectors are usually significantly under loaded and are not suitable for broad band operation. Slightly under loaded projectors can be considered in the same way as electric field limited without a big mistake. Upon substituting expression (3.10) for Q_{mw} and (3.122) for $\dot{W}_{mE}$ into Eq. (3.156) we obtain

$$\Omega_m^2 \approx \left[\frac{E_p^2}{A_E^2}\cdot\frac{r_{ac\,r}}{\dot{W}_r(\Omega_m)} - r_{ac\,r}^2\right]\cdot\frac{1}{(\omega_r M_{eqv})^2}. \qquad (3.157)$$

In the manipulations r_{mL} was neglected in comparison with r_{ac}, because for the electric field limited projectors usually $r_{mL} \ll r_{acr}$. The optimal acoustic load, r_Ω, can be found from this equation for any given value of $\dot{W}_r(\Omega_m)$, which results in the maximum band Ω_m. After differentiating the right side of the equation with respect to r_{ac} and equating the result to zero we arrive at

$$r_\Omega = \frac{1}{2}\left(\frac{E_p}{A_E}\right)^2 \frac{1}{\dot{W}_r(\Omega_m)}. \tag{3.158}$$

The mismatch coefficient

$$m_\Omega = r_{acr}/r_\Omega \tag{3.159}$$

can be introduced to characterize deviation of the real radiation resistance, r_{ac}, from its optimal for broadband operation value. After substituting expression for the optimal load into Eq. (3.157) instead of r_{acr} the following relation will be obtained

$$\Omega_m \dot{W}_r(\Omega_m) = \frac{E_p^2}{2 A_E^2} \cdot \frac{1}{\omega_r M_{eqv}}. \tag{3.160}$$

Thus, the product of the frequency band and the maximum power that can be radiated in this band (gain-bandwidth product) is constant for a particular transducer type under the condition of optimal loading. In general, under an arbitrary acoustic load Eq. (3.160) must be replaced by the inequality

$$\Omega_m \dot{W}_r(\Omega_m) \leq \frac{E_p^2}{2 A_E^2} \frac{1}{\omega_r M_{eqv}}. \tag{3.161}$$

For example, in the case of a cylindrical transducer, which is a typical representative of the projectors with electric field limited maximum acoustic power available for radiation at the resonance frequency, after substituting the corresponding A_E and $\omega_r M_{eqv}$ parameters from Table 3.1 and Table 3.5 we arrive at

$$\frac{\Omega_m \cdot \dot{W}_r(\Omega_m)}{S_\Sigma} \leq 0.5 \cdot \underbrace{E_p^2}_{1} \cdot \underbrace{\left(\frac{d_{3i}}{s_{ii}^E}\right)^2}_{2} \underbrace{\sqrt{\frac{s_{ii}^E}{\rho}} \cdot \frac{t}{a}}_{3}. \tag{3.162}$$

Here $\dot{W}_r(\Omega_m)/S_\Sigma$ is the specific acoustic power radiated (S_Σ is the radiating surface of the projector). There are three factors in the right hand side of the inequality, which essentially affect the capability of the projector to operate in a broad frequency band: the first factor can be identified as the technological factor so far as the value of E_p depends on the transducer fabrication quality; the second factor is the combination of parameters of the piezoelectric material used; the third factor depends on the transducer mechanical system geometry (for the cylindrical transducer this is ratio of the thickness to mean radius of the ring).

The analysis made shows that for considering the operating frequency band of a projector it is necessary to specify a required acoustic power radiated in this band. The acoustic power vs. frequency band tradeoff is illustrated qualitatively by plots in Figure 3.17.

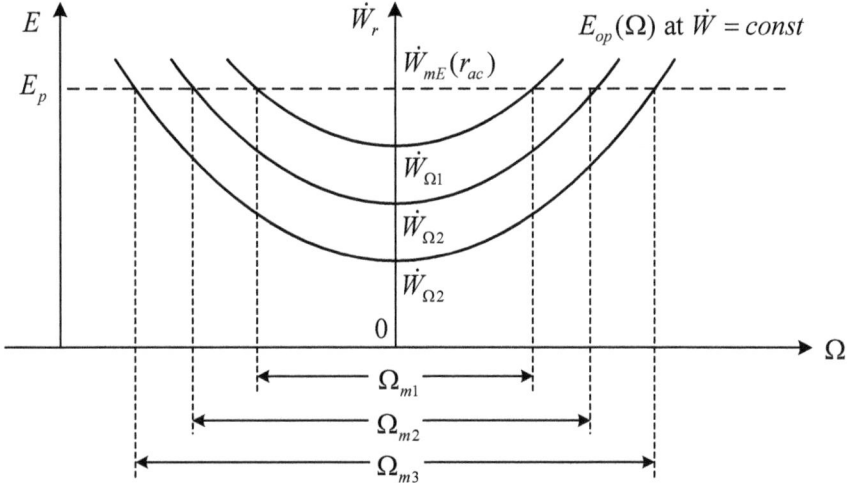

Figure 3.17: Acoustic power available for radiation in a frequency band $\dot{W}_r(\Omega_{mi})$ vs. frequency band Ω_{mi}. Value of the maximum band, in which acoustic power $\dot{W}(\Omega_{mi})$ can be radiated, is on the intersection of its frequency dependence with the line that correspond to the permissible electric field, E_p.

In Figure 3.17 $\dot{W}_r(\Omega_{m1}) > \dot{W}_r(\Omega_{m2}) > \dot{W}_r(\Omega_{m3})$, and $\Omega_{m1} < \Omega_{m2} < \Omega_{m3}$. The product $\Omega_{mi} \cdot \dot{W}_r(\Omega_{mi})$ must not exceed some value that is constant for each transducer type. The maximum power $\dot{W}_{mE}(r_{ac})$ limited by the electric field depends on the radiation resistance and can be radiated at the resonance frequency (at $\Omega = 0$) only. The broader the operating frequency band, the smaller is the acoustic power available for radiation. The maximum operating band, in which a required power can be radiated by a transducer, may be achieved at optimal for this value of power acoustic load given by Eq. (3.158).

The concepts of optimal acoustic loads, under which the maximum values of certain transducer parameters may be achieved, and corresponding mismatch coefficients used in this section need to be emphasized. We have introduced definitions for the optimal loads and mismatch coefficients in terms of the: maximum efficiency at the resonance frequency (r_η by Eq. (3.111) and m_η by Eq. (3.112)); maximum power available at the resonance frequency of a projector (r_W by Eq. (3.126) and m_w by Eq. (3.128)); maximum product $\Omega_m \dot{W}_r(\Omega_m)$ (r_Ω by Eq. (3.158) and m_Ω by Eq. (3.159)). The values of the optimal load resistances are different. Besides, not much freedom is available for actual changing the radiation resistances of projectors. Therefore, the question can arise regarding the usefulness of these concepts. However their importance originates from the fact that the value of an optimal load depends not only on a real radiation resistance, but also on the properties of a particular projector type, as it follows from formulas (3.114), (3.126) and (3.158). The possibility of matching the optimal load by a proper changing the projector design must be considered in each particular case. Thus, the way of optimizing illustrated in this section with example of a cylindrical transducer does not work in the case of a uniform bar, because neither Q_{mw} nor $\dot{W}_r(\Omega)$ depend on the uniform bar geometry. This transducer type is not flexible in terms of optimizing its operation. Under "natural" acoustic loading (in a flat array of a big wave size) projectors of this type are strongly under loaded, and there is no practical way for improving their matching without modifying the mechanical system of the transducer. The well-known and widely used Tonpilz transducer design that employs a mechanical system composed of piezoelectric and passive longitudinally vibrating bars of different lengths and cross sections represents the modification of uniform bar transducer that allows optimizing the acoustic loading. The mass loaded bar transducer considered in Section 2.5 is a variant of this transducer type.

In the conclusion we note that the numerical data related to optimal acoustic loads are presented to illustrate the technique, and they characterize the maximum achievable values, as if the only limiting factor was the electric field. The actual achievable level of power radiated by a transmit channel in a frequency band depends also on the properties of the transducer input impedance and on its effectiveness.

3.2 Transducers in the Receive Mode

3.2.1 Transducer as a Member of Receive Channel

Transducers intended for operating in the receive mode are the sensors that convert energy of a physical input (signal) that characterizes properties of acoustic field (sound pressure, pressure-gradient), or a state of vibration (acceleration) of structures into electrical output. The sensors of the first group are hydrophones and of the second group – accelerometers. The sensors operate as a part of receiving channel, the main task of which is detecting of signals against background noise. In order to formulate the basic properties of the sensors per see, we will consider the receive channel with a single sensor unit. As the background noise will be regarded: the internal noise of the receiving channel that is characterized by value of equivalent noise voltage of the channel reduced to its input; the internal noise of the sensor; and the output voltages of the sensor generated by unwanted actions having a physical nature different from that of the signal (for example, vibration for the hydrophones, and conversely sound pressure for accelerometers). The external actions having the same physical nature as that of the signal, such as the sea noise for hydrophones, will not be considered in this section as a background noise, because it is impossible for a receiving channel with a single sensor to discriminate noise of this kind by changing the sensor design. The only way to reduce the effect of such actions is by means of spatial and temporal data processing that may be performed by an array and signal processing system. The block diagram of a receiving channel is shown in Figure 3.18.

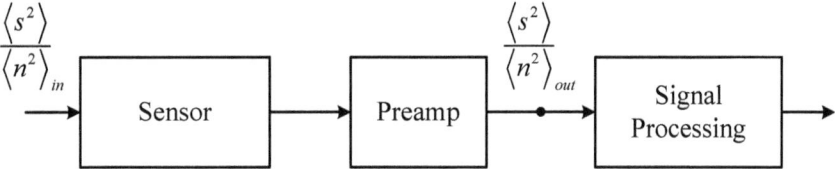

Figure 3.18: Block diagram of a receiving channel.

The sensor and preamplifier make the input of receiving channel. Their goal is to fulfill the "primary processing" of a signal, i.e., to raise magnitude of the signal delivered to the "secondary processing" electronics to a level sufficient to avoid a deterioration of the signal to noise ratio in the subsequent circuits. The problem is in minimizing the signal to noise decrease in process of the primary processing, as both the sensor and preamplifier are the sources of additional noise.

In terms of reducing influence of combined internal noise of the sensor itself and of the receiving channel on the signal to noise ratio, the properties of the sensor as the source of energy of a signal are important, because the minimal signal must produce an output effect exceeding level of the preamplifier noise to a certain degree. These properties are considered in Section 3.2.2. An important parameter of a sensor regarding its internal noise is the threshold signal, which is defined by the value of the minimal external action that produces the output voltage exceeding the sensor internal noise level. These issues are considered in Section 3.2.3.

In terms of diminishing influences of the external unwanted actions, the noise immunity of the sensors, as their ability to be as less sensitive to these actions, as it is needed under the particular operating conditions, will be considered in Section 3.2.4. It is noteworthy that requirements for different parameters of the sensors may be more or less challenging depending on their applications. Thus, the most demanding are requirements for the sensors intended for populating arrays of the passive sonars. For measurement sensors that usually deal with strong signals more important may be requirements for their immunity to unwanted actions.

3.2.2 Sensor as a Source of Energy for the Receive Channel

In accordance with the Thevenin's theorem, a sensor can be considered as a source of energy for the receiving channel with the electromotive force equal to the output voltage of the open circuited sensor (V_{oc} in the equivalent circuit in Figure 1.18 that is reproduced here as Figure 3.19) and with internal impedance Z_{int} equal to the impedance of the circuit between points 1, 1.

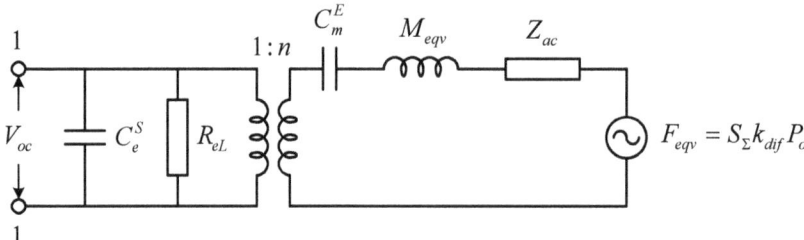

Figure 3.19: Equivalent circuit of a sensor.

3.2. Transducers in the Receive Mode

The open circuit voltage for a hydrophones can be represented from the equivalent circuit (at the condition $\omega C_e^S R_{eL} \gg 1$ that usually holds) as

$$V_{oc} \approx \frac{1}{j\omega C_e^S} \cdot \frac{nF_{eqv}}{Z_{me}^E} = \frac{1}{j\omega C_e^S} \cdot \frac{nS_\Sigma k_{dif}(\omega, r)}{Z_{me}^E} P_0 = \gamma_{oc}(\omega, r) \cdot P_0, \qquad (3.163)$$

where Z_{me}^E and F_{eqv} are determined by relations (1.75) and (2.30); P_0 is the sound pressure in the plane wave propagating in the free field in direction of the acoustic axis of the hydrophone and $\gamma_{oc}(\omega, r)$ is the free field open circuit sensitivity of the hydrophone. The sensitivity of a hydrophone depends on the direction of propagating the plane wave in accordance with directional factor in the same way as diffraction coefficient for the transmit mode (2.38) depends. Further its value $\gamma_{oc}(\omega)$ will be considered that corresponds to direction of the acoustic axis,

$$\gamma_{oc}(\omega) = \frac{1}{j\omega C_e^S} \cdot \frac{nS_\Sigma k_{dif}(\omega)}{Z_{me}^E}. \qquad (3.164)$$

The open circuit sensitivity, OCVS, (further just sensitivity for brevity) being expressed in the dB scale is called RVS – receive voltage response, or sometimes FFVS – free field voltage sensitivity,

$$\text{OCVS} = 20\log \gamma_{oc} \text{ re } 1 \text{ V/}\mu\text{Pa}. \qquad (3.165)$$

Requirements for the sensitivity of a hydrophone are quite different for its operation in the frequency band around the resonance frequency and at frequencies below the resonance frequency. Operation in the band around the resonance frequency is typical for the reversible mode of transducer operation, in which case the requirements for transmit mode of operation dominate, and requirements for the receive mode are met automatically. In the case that phase relations between transmitted and received signals are important it must be remembered that the resonance frequency and hence the phase characteristic of a transducer in the transmit and receive modes are different. Whereas the condition for resonance frequency in the transmit mode is $\text{Im}\{Z_m^E = 0\}$ and $\omega_{rw} = 1/\sqrt{(M_{eqv} + m_{ac})C_{eqv}^E} = \omega_{ra}/\sqrt{1+(x_{ac}/\omega M_{eqv})}$ (see examples in Table 3.1), the condition for resonance frequency in the receive mode is $\text{Im}\{Z_{me}^E = 0\}$. Following relation (1.75) it can be obtained that

$$\omega_{rme} = 1/\sqrt{(M_{eqv} + m_{ac})C_{eqv}^E / (1+\alpha_c)}, \qquad (3.166)$$

where $\alpha_c = n^2 C_{eqv}^E / C_e^S$ and $(1+\alpha_c) = 1/(1-k_{eff}^2)$ according to relations (2.93). Thus,

$$\omega_{me\,w} = \omega_{ar\,a} / \sqrt{1 + (x_{ac} / \omega M_{eqv})} \,. \tag{3.167}$$

Here $\omega_{ar\,a}$ is the antiresonance frequency of the transducer in air.

Comparison is made in Figure 3.20 between the frequency responses of a spherical transducer in the transmit and receive modes normalized to their values at the resonance frequencies.

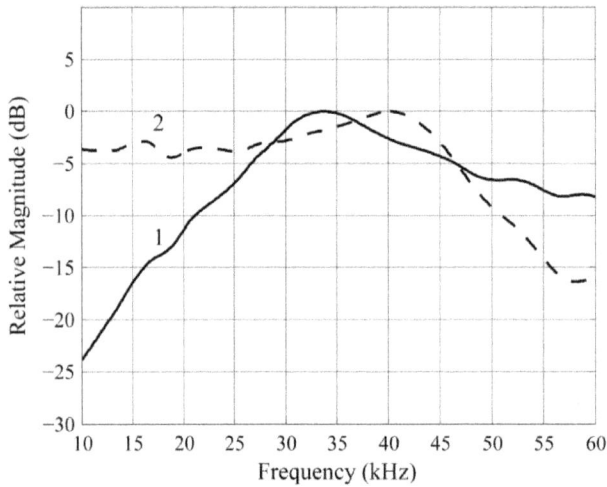

Figure 3.20: Comparison between the frequency responses of a spherical transducer at $a/t = 9.5$ in (1) transmit and (2) receive modes normalized to their magnitude at the peak response.

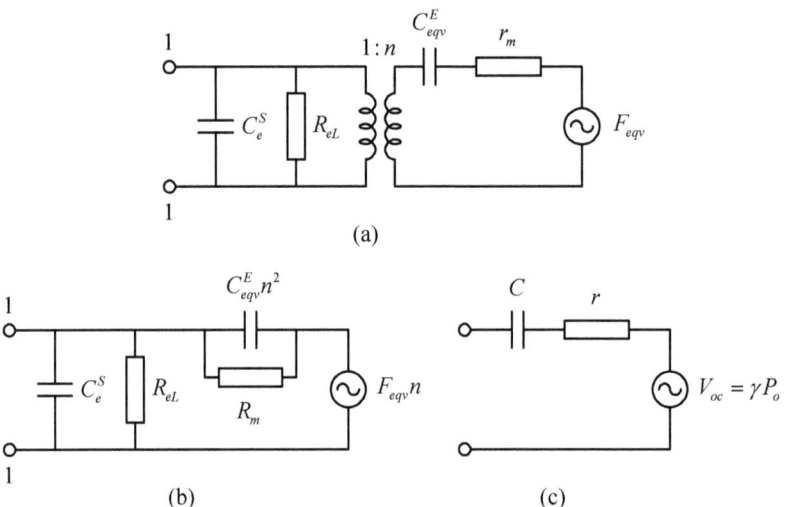

Figure 3.21: (a) Equivalent circuit of a sensor operating at frequencies much below the resonance ж(b) transformation of the circuit to the electrical side (c representation of the sensor as equivalent electrical generator for the receive channel

An important characteristic of the measurement hydrophones is uncorrupted signal reproduction in a broad frequency range. One of the basic requirements for the sensors intended for application in arrays of passive sonars is a small spread in performance (magnitude and phase identity). Therefore, the most typical mode of operation for these groups of sensors is in the frequency range below resonance, where frequency response of a sensor is relatively flat. The frequency range below resonance is characterized by the fact that in the equivalent circuit of a transducer $\omega M_{eqv} \ll 1/\omega C_{eqv}^E$. Equivalent circuit of the sensor operating at frequencies below resonance is shown in Figure 3.21 together with its representation as the equivalent generator. Expression for the parameters of the equivalent generator in Figure 3.21 (c) at low frequencies greatly simplifies. At $\omega \ll \omega_r$

$$Z_{me}^E \approx \frac{1}{j\omega C_{eqv}^E} + \frac{n^2}{j\omega C_e^E} = \frac{1+\alpha_c}{j\omega C_{eqv}^E} = \frac{1}{j\omega C_{eqv}^E (1-k_{eff}^2)}, \qquad (3.168)$$

where expressions (2.93) for coefficients α_c and k_{eff}^2 are considered. Besides, it follows from Figure 3.21 (a) that capacitance C_{Lf} measured between terminals 1, 1 is

$$C_{Lf} = C_e^E + n^2 C_{eqv}^E = \frac{C_e^E}{1-k_{eff}^2}. \qquad (3.169)$$

Thus, expression (3.164) for the open circuit sensitivity becomes

$$\gamma_{oc}(\omega) = \frac{C_{eqv}^E}{C_{Lf}} \cdot nS_\Sigma k_{dif}(\omega). \qquad (3.170)$$

The active resistances are neglected in this calculation because they are insignificant compared with the reactive terms. The internal impedance $Z_{in\,1,1}$ of the generator must be calculated as impedance between the terminals 1, 1 at the condition that the voltage source inside the circuit (F_{eqv} is short circuited (see definition for the generalized generator done in regard to Figure 1.16).

The resistances of losses cannot be neglected in determining the internal impedance though being small compared with the reactive terms, because they constitute the source of the internal noise of the sensor. The magnitude of resistance of mechanical losses, r_{mL}, ideally can be expressed through the mechanical quality of piezoceramic material used, as $r_{mL} = 1/\omega C_{eqv}^E Q_m$. This quantity may increase in a finished transducer design. We will assume that this increase is negligible to the first approximation. The radiation resistance of a single transducer, r_{ac}, depends in general on the transducer type and wave size. Examples of estimating radiation resistances for different transducer types are given in Table 3.1. In the case that sensors operate in arrays, their radiation resistances may depend on configuration of an array and of a relative position of the sensors due to acoustic interaction between them. Resistance R_m in Figure 3.21 (b) can be represented as

$$R_m = Q_{m\Sigma} n^2 / \omega C_{eqv}^E, \tag{3.171}$$

where

$$Q_{m\Sigma} = 1 / \omega C_{eqv}^E r_m \text{ and } r_m = r_{mL} + r_{ac}. \tag{3.172}$$

All of these quantitatives have to be determined by analizing the elements of equivalent circuit in Figure 3.21 (a). The dielectric loss factor of a finished sensor design also may differ from analogous parameter of the piezoceramics, but for carefully designed sensor this difference can be small enough. At least it will be neglected in the context of this Section, and the resistance of electrical losses, R_{eL}, will be represented as $R_{eL} = 1 / \omega C_e^S \tan \delta_e$.

Finally, the sensor may be represented as a source of energy for the preamplifier by the circuit in Figure 3.21 (c). The internal impedance of the sensor can be considered as purely capacitive with capacitance C_{Lf} in calculating the signal transfer. And the resulting resistance of the internal losses that determines property of the sensor as a source of internal noise is

$$r_{in} = \frac{1}{\omega C_{Lf}} \left[(1 - k_{eff}^2) \tan \delta_e + \frac{k_{eff}^2}{Q_{m\Sigma}} \right]. \tag{3.173}$$

In order to characterize this property, the electrical quality factor of a sensor can be introduced as

$$Q_{tr} = \frac{1}{\omega C_{Lf} r_{in}} = \frac{1}{\tan \delta_{tr}}, \tag{3.174}$$

where

$$\tan \delta_{tr} = (1 - k_{eff}^2) \tan \delta_e + \frac{k_{eff}^2}{Q_{m\Sigma}}. \tag{3.175}$$

These quantities can be useful also in terms of experimental estimating of internal losses. It is instructive to compare contribution of electrical and mechanical losses to the total loss effect. Determine the quality factor $Q_{m\Sigma}$ for the typical low frequency double-sided flexural plate sensor design shown in Figure 2.16. After substituting expressions (2.151) for $C_{eqv}^E = 1/K_{eqv}^E$ and (2.153) for r_{ac} (given that the wave size of the sensor is $ka < 0.6$) in formula (3.172) will be obtained

$$\frac{1}{Q_{m\Sigma}} = \tan\delta_m + 37\frac{(\rho c^2)_w}{(\rho c^2)_c}\frac{a^6}{\lambda^3 t^3}, \qquad (3.176)$$

where $\tan\delta_m$ and $(\rho c^2)_c$ are the quantities determined for a ceramic composition, and $(\rho c^2)_w$ is for water, respectively. From the condition that $ka < 0.6$ follows $a/\lambda < 0.1$. The aspect ratio a/t for the sensors operating under moderate hydrostatic pressures can be about 5 to 7, so let be $(a/t) < 7$. Thus, for the sensor built from PZT-4 ceramics

$$\frac{1}{Q_{m\Sigma}} \approx \tan\delta_m + 0.35. \qquad (3.177)$$

Using relation (3.175), we arrive at following estimation of ratio of mechanical to electrical components of the loss factor $\tan\delta_{tr}$

$$\frac{(k_{eff}^2/Q_{m\Sigma})}{(1-k_{eff}^2)\tan\delta_e} \approx \frac{k_{eff}^2}{1-k_{eff}^2}\frac{\tan\delta_m + 0.35}{\tan\delta_e}. \qquad (3.178)$$

Conservatively, $\tan\delta_m < 0.01$ and can be neglected. Expression for the effective coupling coefficient (2.152) for PZT-4 results in $k_{eff}^2 = 0.18$. Thus, contribution of the mechanical (acoustic) losses may be significant. The ratio (3.178) is 1.6 in this particular example, given that for PZT-4 $\tan\delta_e \approx 0.05$.

The open circuit sensitivity and low frequency capacitance will be denoted further for simplicity without subscripts, namely, γ and C instead of γ_{oc} and C_{Lf}. These parameters fully characterize a sensor as the source of signal. However, it is hard to judge sensors by using these parameters alone. Therefore, it is desirable to establish Figure of Merit for a sensor as the source of signal. This figure of merit can be derived from the obvious consideration that out of two sensors having equal internal impedances, the sensor that has larger sensitivity is better. Consider two sensors with different sensitivities (γ_1, γ_2) and different internal impedances (X_1, X_2) loaded with the same impedance Z_L, as represented by Figure 3.22 (a) and (b).

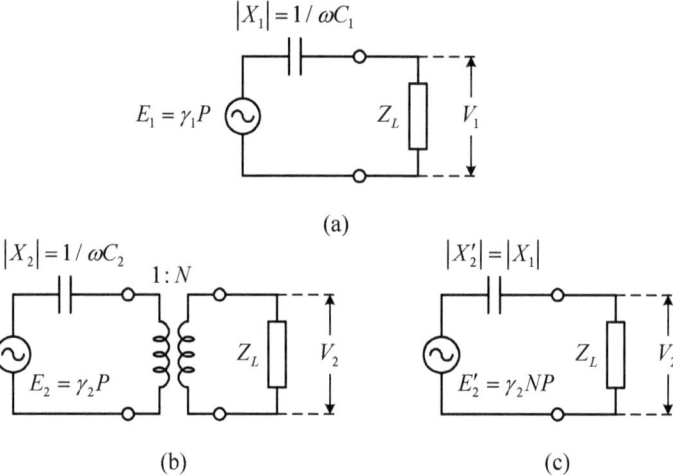

Figure 3.22: Comparison of outputs of the sensors loaded with the same impedance Z_L: (a) equivalent circuit of sensor number 1; (b) circuit of sensor number 2 having different sensitivity and internal impedance, and ideal transformer; (c) circuit of sensor number 2 after transformation.

Let us assume that the load is applied to the sensor number 2 via an ideal transformer with turn's ratio $N = \sqrt{X_1/X_2}$ as illustrated in Figure 3.22 (b). Parameters of the sensor number 2 after the transformation become $\gamma_2' = \gamma_2 N$ and $|X_2'| = |X_2| \cdot N^2 = |X_1|$, as shown in Figure 3.22 (c). Comparing output voltages of the sensors V_1 and V_2 results in

$$\frac{V_1}{V_2} = \frac{E_1}{E_2} = \frac{\gamma_1}{\gamma_2 N} = \frac{\gamma_1 \sqrt{C_1}}{\gamma_2 \sqrt{C_2}}, \qquad (3.179)$$

which means that the sensor having the larger parameter $\gamma \sqrt{C}$ is potentially better as the source of signal. We will define this parameter as the specific sensitivity,

$$\gamma_{sp} = \gamma \sqrt{C}. \qquad (3.180)$$

The specific sensitivity does not change by using an ideal transformer. In practice, the performance of an ideal transformer can be duplicated by a proper designing of the sensor. This is illustrated with the examples shown in Figure 3.23. As it can be seen from Figure 3.23, the parallel to series switching of the piezoelements (Figure 3.23 (a)), or single sensor units in the case that the sensor is comprised of a number of separate units (Figure 3.23 (b)) the sensor is comprised of a number of separate units (Figure 3.23 (b)) is analogous to an ideal transformation. Further transformation can be achieved by dividing electrodes of a single piezoelement in several parts and switching them to series connection (Figure 3.23 (c)). Such a redesigning of the electrodes is advantageous for achieving increased sensitivity and matching with the preamplifier, as it will be illustrated in Section 3.2.3.

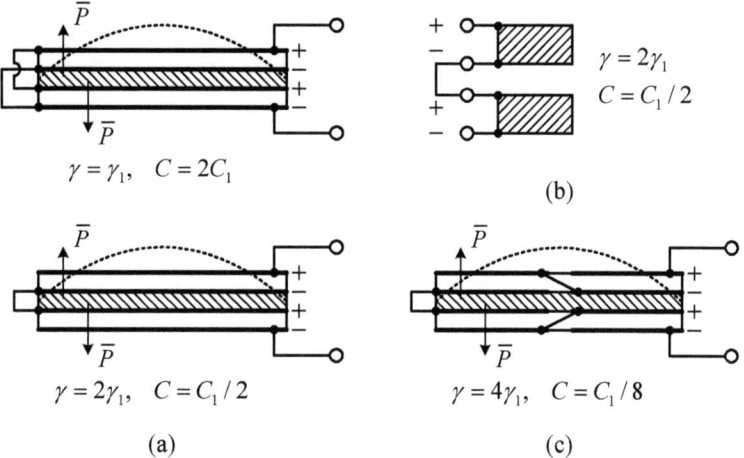

Figure 3.23: Transformation of the output signals: (a) by switching electrodes of a piezoelement from parallel to series; (b) by changing connection of the single transducer units, (c) by dividing electrodes of the piezoelements into parts and connecting the parts in series. The sensitivity and capacitance of a single piezoelement (single sensor unit) are γ_1 and C_1. In all the cases

$$\gamma\sqrt{C} = \gamma_1\sqrt{C_1}$$

Using relations (3.163) and (3.170) for γ and C and definition for k_{eff} (2.93), the expression for γ_{sp} can be obtained

$$\gamma_{sp} = k_{eff} S_{av} k_{dif} \sqrt{C_{eqv}^E} . \qquad (3.181)$$

(Note that generally for low frequency sound pressure hydrophones $k_{dif} = 1$, but for the pressure gradient hydrophones it should be determined by formula (2.162)). The value γ_{sp} characterizes the rated power of the sensor as the energy of a signal source. Indeed, when the latter operates under the matched electrical load $R_{el.l} = 1/\omega C_{el}$,

$$\dot{W}_{rt} = \omega \frac{[\gamma\sqrt{C}]^2}{4} |p_o|^2 . \qquad (3.182)$$

If the overall surface area S_Σ occupied by the hydrophone is limited (as it can be in case of populating an array), the more objective criterion for sensor can be quantity

$$\gamma_{sp} / \sqrt{S_\Sigma} = \gamma_{rd}, \qquad (3.183)$$

which will be called the reduced specific sensitivity. The reduced sensitivity characterizes the efficiency of the acoustoelectric conversion produced by a sensor per unit area. This becomes clear, if to relate $\dot{W}_{rt}$ to value of the acoustic energy flux $\dot{W}_{ac}$ that passes in the free field through the "dimensional area" S_Σ occupied by the sensor, $\dot{W}_{ac} = (|p_0|^2 / 2\rho c) S_\Sigma$. This will result in

$$\frac{\dot{W}_{rt}}{\dot{W}_{ac}} = \omega \frac{(\rho c)_w}{2} \frac{\gamma_{sp}^2}{S_\Sigma} = \omega \frac{(\rho c)_w}{2} \gamma_{rd}^2. \qquad (3.184)$$

While γ_{sp} can be increased by increasing the sensor dimensions, which do not affect the resonance frequency, such as the cylinder height or end area of a bar, or by using several electrically connected identical sensor units, the reduced sensitivity remains the same. It belongs to a sensor type. The achievable values of sensitivities of the hydrophones essentially depend on the range, in which their frequency response must remain linear, and on the depth of operation. These issues will be considered in Chapter 14.

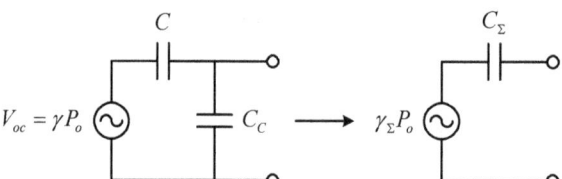

Figure 3.24: Transformation of the sensor as equivalent generator circuit in case that it is ended by a cable with capacitance C_C.

In practice, sensor may be connected to preamplifier by a cable having capacitance C_c that depends on its length. In this case the equivalent generator representing the sensor must be transformed, as it is shown in Figure 3.24. In the transformed circuit

$$\gamma_\Sigma = \gamma \frac{C}{C + C_c} \text{ and } C_\Sigma = C + C_c. \qquad (3.185)$$

3.2. Transducers in the Receive Mode

3.2.3 Noise Property of a Receive Channel, Requirement for the Sensor Sensitivity.

3.2.3.1 Internal Noise of a Sensor

The internal resistance (3.173) is the source of thermal self-noise of the sensor. In the representation of the sensor as equivalent generator it can be replaced by the equivalent noise voltage e_n with mean square value per 1 Hz bandwidth.

$$\langle e_n^2 \rangle = 4k_B Tr = 4k_B T \frac{\tan \delta_{tr}}{\omega C}, \tag{3.186}$$

where k_B is Boltzmann's constant, $k_B = 1.38 \cdot 10^{-23} \, J/°K$, T is the absolute temperature and the brackets $\langle \, \rangle$ indicate time average. Finally, the electrical circuit of the sensor as a source of signal and of internal noise may be represented as shown in Figure 3.25.

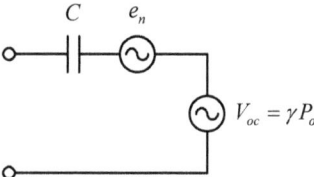

Figure 3.25: Electrical circuit of a sensor as source of signal and noise.

The figure of merit for a sensor as a source of noise can be introduced by considering the minimum detectable signal, p_{min}, as the signal having such a magnitude that its distortion due to the sensor internal noise is limited to a certain permissible extent. In other words, the minimum signal to internal noise ratio at the sensor output must be of a certain permissible level β_1, or

$$\frac{p_{min} \gamma}{e_n} = p_{min} \sqrt{\frac{\omega}{4k_B T}} \cdot \frac{\gamma \sqrt{C}}{\sqrt{\tan \delta_{tr}}} \geq \beta_1, \tag{3.187}$$

where from

$$p_{min} \geq \beta_1 \sqrt{\frac{4k_B T}{\omega}} \cdot \frac{\sqrt{\tan \delta_{tr}}}{\gamma_{sp}}. \tag{3.188}$$

The value of coefficient β_1 depends on the signal processing properties of the receive system. The figure of merit in expression (3.188) that characterizes the sensor as a contributor of noise is the quantity $\sqrt{\tan \delta_{tr}}/\gamma_{sp}$ (the smaller this quantity, the better). Thus, in order to minimize the detectable signal, the specific sensitivity of the sensor should be maximized in addition to obvious requirement of minimizing internal noise by reducing mechanical and electrical losses to the level that is inherent in the ceramic composition used.

3.2.3.2 Matching a Sensor with Preamplifier

Consider a sensor as a part of the receiving system having block diagram that is illustrated in Figure 3.18. As it was noted, the goal of the system is to detect the acoustic signals existing in a background of ambient sea noise. The goal of the sensor and preamplifier as part of the system is in minimizing degradation of signal to noise ratio at output of the preamplifier in comparison with this ratio at the input of the sensor in course of the primary processing of the signals. The relative decrease of the signal to noise ratio can be characterized by the relation

$$\frac{\langle s^2 \rangle / \langle n^2 \rangle_{out} - \langle s^2 \rangle / \langle n^2 \rangle_{in}}{\langle s^2 \rangle / \langle n^2 \rangle_{in}} \leq \beta, \qquad (3.189)$$

where $\langle s^2 \rangle / \langle n^2 \rangle_{out}$ and $\langle s^2 \rangle / \langle n^2 \rangle_{in}$ are the signal to noise power ratios at preamplifier output and in the sound field, accordingly, and β is the measure of permissible degradation. In order to derive requirements for the sensor parameters from relation (3.189), the primary processing part of the receiving system has to be considered that includes the sensor as generator of signal and internal noise, and preamplifier, which produces the gain and also contributes to a noise level. The equivalent circuit of this part of the receiving channel is shown in Figure 3.26.

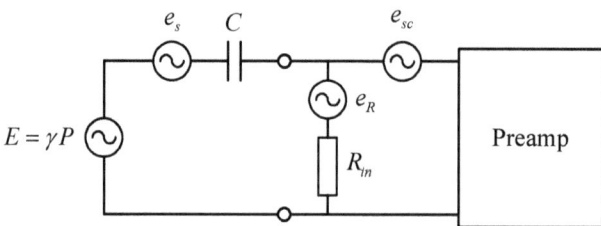

Figure 3.26: Equivalent circuit of input of the receiving channel with sources of internal noise.

The sensor is represented by the circuit in Figure 3.25. The preamplifier is represented as a noiseless gain-producing block with sources of internal noise referred to its input. It is assumed that the sensor is connected to the amplifier directly without a cable. The noise contribution from the amplifier is represented in this circuit by two generators e_{sc} and e_R, which can

3.2. Transducers in the Receive Mode

be considered uncorrelated. The e_{sc} represents voltage noise of the amplifier with the input short-circuited. And e_R is the thermal noise of an equivalent input resistance of the amplifier, R_{in}. Its mean square value per 1 Hz bandwidth is

$$\langle e_R^2 \rangle = 4k_B T R_{in}. \tag{3.190}$$

The amplifier noise parameters can also be considered as S_e, the spectral density of the voltage source, S_e, and the spectral density of the current source, S_i. In our notation

$$S_e = \langle e_{sc}^2 \rangle, \quad S_i = \langle e_R^2 \rangle / R_{in}^2 = 4k_B T / R_{in}. \tag{3.191}$$

We will not consider the situation in which sensor is connected to preamplifier via cable, by the two reasons. Primarily, in an optimal designed receiving system the best result can be obtained in the case that preamplifier is near the sensor, if not combined with the sensor design. Moreover, the modern state of electronic device design allows this to be done practically without compromising the system's reliability. Secondly, existing of a cable can be taken into consideration by changing the sensor parameters according to relations (3.185). We will assume that to ensure that the low frequency roll-off is well below the operating frequency band the relation holds

$$(\omega_L C R_{in})^2 \gg 1 \tag{3.192}$$

(for example, $\omega_L C R_{in} > 3$), where ω_L is the lowest operating frequency. In this case the circuit in Figure 3.25 can be modified, as it is shown in Figure 3.27. In Figure 3.27 $r = 1/(\omega C)^2 R_{in}$, and e_r is the thermal noise of resistance r with the spectral density

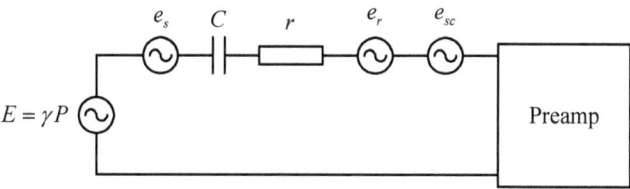

Figure 3.27: Modified circuit of the input of receiving system.

$$\langle e_r^2 \rangle = 4k_B T / (\omega C)^2 R_{in}. \tag{3.193}$$

The resistance r may be neglected due to relation (3.192).

The signal to noise ratio at the omnidirectional receiving system input (with a single sensor of small wave size) is

$$\langle s^2 \rangle / \langle n^2 \rangle_{in} = \langle p_s^2 \rangle / \langle p_{sn}^2 \rangle. \qquad (3.194)$$

where p_s is the signal pressure and p_{sn} is the ambient sea noise pressure measured by the omnidirectional hydrophone that is specified as background noise for the system. All the sources of noise and signal at the amplifier input are uncorrelated, therefore their energies may be summed and the signal to noise ratio at the amplifier output will be

$$\left.\frac{\langle s^2 \rangle}{\langle n^2 \rangle}\right|_{out} = \frac{\langle p_s^2 \rangle \gamma^2}{\langle p_{sn}^2 \rangle \gamma^2 + \langle e_n^2 \rangle + \langle e_r^2 \rangle + \langle e_{sc}^2 \rangle} = \frac{\langle p_s^2 \rangle}{\langle p_{sn}^2 \rangle} \bigg/ \left[1 + \frac{\langle e_n^2 \rangle + \langle e_r^2 \rangle + \langle e_{sc}^2 \rangle}{\langle p_{sn}^2 \rangle \gamma^2}\right]. \qquad (3.195)$$

Combining the expressions (3.189), (3.194) and (3.195) we arrive at relation

$$\frac{\langle e_n^2 \rangle + \langle e_r^2 \rangle + \langle e_{sc}^2 \rangle}{\langle p_{sn}^2 \rangle \gamma^2} \leq \beta. \qquad (3.196)$$

Thus, the ratio of internal noise of the receiving system to the noise generated by the ambient sea noise pressure at the preamplifier input should be less than the coefficient β, which is the measure of the acceptable degradation of the signal to noise ratio. Following relation (3.196), the minimum sensitivity required to fulfill this condition is

$$\gamma^2 \geq \frac{\langle e_n^2 \rangle + \langle e_r^2 \rangle + \langle e_{sc}^2 \rangle}{\beta \langle p_{sn}^2 \rangle}. \qquad (3.197)$$

Upon substituting $\langle e_n^2 \rangle$ and $\langle e_r^2 \rangle$ by their expressions (3.186), (3.193) and taking into account the expression (3.180) for γ_{sp}, inequality (3.197) will be transformed into

$$\gamma_{sp}^2 \geq \frac{\langle e_{sc}^2 \rangle \omega C + 4k_B T / \omega C R_{in} + 4k_B T \tan \delta_{tr}}{\beta \omega \langle p_{sn}^2 \rangle}. \qquad (3.198)$$

The numerator in expression (3.198) depends on the sensor capacitance C that can be changed in the sensor design, whereas the specific sensitivity γ_{sp} will remain intact. Thus, the optimum value of the capacitance can be found that matches an amplifier in terms of minimizing the required sensor sensitivity. This value may be obtained by differentiating the numerator with respect to C and equating the result to zero. After performing the differentiation, the optimum value of the capacitance will be found, as

$$C_{opt} = \frac{\sqrt{4k_B T}}{\omega\sqrt{\langle e_{sc}^2 \rangle R_{in}}}. \tag{3.199}$$

This relation may be represented alternatively in terms of the spectral densities S_e and S_i of the amplifier noise using the expressions (3.191), as

$$C_{opt} = \frac{1}{\omega} \cdot \sqrt{S_i / S_e}. \tag{3.200}$$

Analogous result in terms of matching sensor with preamplifier was demonstrated in Refs. 12 and 13 using a different approach. The condition (3.192) for the low frequency roll-off should be verified at $C = C_{opt}$ based on the relation

$$(\omega C_{opt} R_{in})^2 = (4k_B T)^2 / S_i S_e. \tag{3.201}$$

With C_{opt} known the minimum specific sensitivity required to meet relation (3.198) can be found. Upon substituting expressions (3.199) or (3.200) for C_{opt} into relation (3.198) one may arrive at the following results

$$(\gamma_{sp}^2)_{min} = \frac{1}{\omega \beta \langle p_{sn}^2 \rangle} \left(2\frac{\sqrt{4k_B T \langle e_{sc}^2 \rangle}}{\sqrt{R_{in}}} + 4k_B T \tan \delta_{tr} \right) \tag{3.202}$$

or

$$(\gamma_{sp}^2)_{min} = \frac{1}{\omega \beta \langle p_{sn}^2 \rangle} \left(2\sqrt{S_i S_e} + 4k_B T \tan \delta_{tr} \right). \tag{3.203}$$

In both relations (3.202) and (3.203) the first term within the bracket is due to amplifier noise and the second term is due to sensor internal noise, where $\tan \delta_{tr}$ is determined by relation (3.175).

3.2.4 Response of the Sensors to Unwanted Actions

Under operating environmental conditions, the sensors may be exposed to the unwanted actions having a physical nature other than that of a signal, but also resulting in generating electric voltage at their output, which can be considered as noise. If to denote the sensor sensitivity to an unwanted action as γ_N and the action itself as F_N, then the output voltage generated by this action is $V_N = \gamma_N F_N$. In evaluating an effect of the unwanted action on the sensor operation it is not the absolute value of the output voltage that matters, but its relation to the value of voltage generated by signal $V_s = \gamma_s F_s$, namely, signal to noise ratio

$$V_s/V_N = (\gamma_s/\gamma_N)(F_s/F_N) = NI \cdot (F_s/F_N). \qquad (3.204)$$

We define the parameter $NI = \gamma_s/\gamma_N$ that depends on the sensor design and determines its immunity to an unwanted action, as the coefficient of noise immunity (*NI*). The ratio F_s/F_N characterizes the external conditions, under which reception of signal takes place. The value of this ratio determines the requirements for the sensor immunity to a particular unwanted action.

Thus, a sensor intended for measuring the sound pressure (zero-order sensor) may experience in operating conditions an action caused by the structural vibrations propagating through its mounting elements, or/and by the pressure gradient in the sound field. For a sensor of the first-order in addition to structural vibrations the sound pressure is an unwanted action, to which an ideal sensor must not react by definition. For accelerometers, designated for measuring a single component of acceleration of bodies and surfaces under the real operating conditions, the unwanted actions are the components of vibration in the perpendicular directions and the sound pressure that can be generated by the vibrating body itself or by an independent source. The sound pressure may produce deformations of the accelerometer case, which in its turn may cause deformations of the accelerometer piezoelement and generating the unwanted output voltage.

The concept of sensor noise immunity will be illustrated in this section with example of pressure (zero order) hydrophone immunity to vibration. Detailed analysis regarding noise immunity of sensors of different kind and the ways of its increasing will be done in Chapter 14. The following notations will be introduced when considering the noise immunity of pressure hydrophone to vibration. The output voltage and sensitivity of the hydrophone related to action of the sound pressure will be denoted by subscript *p* that corresponds to the nature of the signal, i.e., V_p and γ_p. The output voltage and sensitivity of the hydrophone to unwanted action (acceleration) will be denoted with two indices: with the subscript *p* that corresponds to the signal,

3.2. Transducers in the Receive Mode

and with superscript $\dot{U}$ that corresponds to the unwanted action of the acceleration, i. e., V_p^U and γ_p^U. By definition of the sensitivity $V_p = \gamma_p p$ and $V_p^U = \gamma_p^U \dot{U}$. Thus, relation (3.204) becomes

$$\frac{V_s}{V_N} = \frac{V_p}{V_p^U} = \frac{\gamma_p}{\gamma_p^U} \frac{P}{\dot{U}} = NI_p^U \frac{P}{\dot{U}}. \qquad (3.205)$$

Here, $NI_p^U = \gamma_p / \gamma_p^U$ is the coefficient of immunity of sound pressure hydrophone to acceleration that otherwise can be called the "vibration resistance" of the hydrophone.

It is noteworthy that the sound pressure hydrophone even of small wave size does not remain omnidirectional, if it is sensitive to acceleration, because the component of its output generated by the acceleration is $V_p^U = \gamma_p^U \dot{U} \cos\theta$, where θ is the angle relative to direction of the acceleration (see Eqs. (2.125) and (2.127)). When estimating the noise immunity, the maximum sensitivity to the vibration must be considered unless the direction of its action is known. Thus, in order to qualitatively estimate the noise immunity we will use relation (3.205), where γ_p^U is the maximum value of sensitivity to vibration.

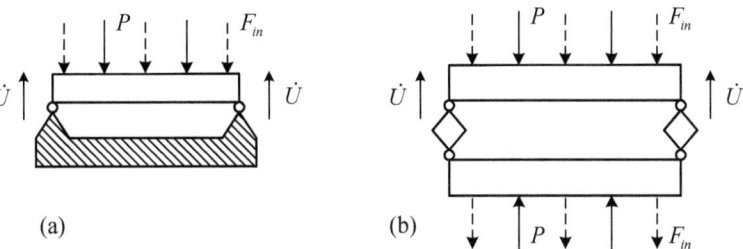

(a)   (b)

Figure 3.28: Examples of the single plate (a) and symmetrical double plate (b) hydrophones that are subjected to vibration with acceleration $\dot{U}$. Shaded is the passive plate. Solid arrows show direction of acting of sound pressure, dashed arrows show direction of action of inertia forces due to the acceleration.

To illustrate, how the vibration resistance can depend on the sensor design, consider examples of the flexural plate sound pressure hydrophones in the variants of single plate and symmetrical double plate designs that are shown in Figure 3.28. Assuming that the piezoceramic plates comprising the hydrophones are identical, the sensitivities of the hydrophones can be compared by values of the equivalent forces acting at their surfaces.

For the single plate hydrophone they are: $F_{eqv\,p} = S_{av}P$ due to the sound pressure, and $F_{eqv\,\dot{U}} = \rho t S_{av}\dot{U}$ due to the acceleration (see formula (2.148)). Thus,

$$NI_p^{\dot{U}} = 1/\rho t . \qquad (3.206)$$

For the symmetrical double plate hydrophone $F_{eqv\,p} = 2S_{av}P$, and $F_{eqv\,\dot{U}} = 0$ in the case that the plates are electromechanically and mechanically identical, because the output voltages generated by the acceleration in the plates have equal magnitudes and opposite phase. Thus, theoretically the double plate hydrophone is ideally acceleration resistant. Practically, some difference in parameters of the plates may exist, and the noise immunity coefficient $NI_p^{\dot{U}} \neq 0$. If value of this coefficient is larger than admissible, it can be further reduced by equalizing the output voltages generated in the plates by acceleration. A procedure for equalizing the sensitivities may be accomplished by external tuning as illustrated in Figure 3.29. By changing the capacitance C_{ad} connected in parallel to the plate having larger sensitivity, the output voltage V_{out} can be reduced theoretically to zero (and practically to the noise level of instrumentation used). This will be reached at

$$C_{ad} = C_1(\gamma_1 - \gamma_2)/\gamma_1 . \qquad (3.207)$$

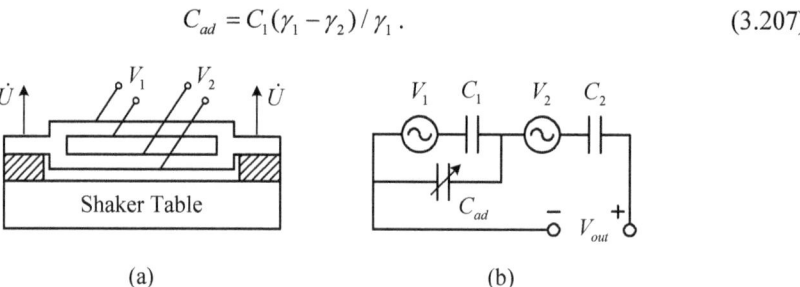

Figure 3.29: Illustration of method for equalizing the sensitivities of individual plates by external capacitive tuning.

As one more example, consider the structural noise immunity of the widely used broadband hydrophones that employ cylindrical piezoelements operating in a frequency range much below their resonance frequency. The acceleration resistance of the hydrophones depends significantly on the way, how the cylindrical piezoelement is attached to a vibrating structure. In the case that cylindrical piezoelement is attached to the supporting structure by its end, as it is shown in Figure 3.30 (a), vibration with acceleration $\dot{U}$ propagating through the structure in the direction of the cylinder axis will generate the inertia forces, and corresponding mechanical stress in the piezoelement will be $T(x) = \dot{U}\rho(l-x)$ (l is the length of a cylinder and ρ is the density of its material). It is easy to verify using results of Section 2.3 that the ratio of the output voltages of the piezoelement caused by the sound pressure and by the inertial forces due to acceleration is $V_p / V_p^U = (P/\dot{U})(2/tl\rho)$, where a is the mean radius of the cylinder and t is its thickness. Thus, the vibration resistance of the cylinder in the axial direction has a finite value

$$NI_p^U = 2/tl\rho. \qquad (3.208)$$

Sensitivity of the cylindrical piezoelement to vibration in the radial direction must vanish due to symmetry of the cylinder cross section, assuming that it is mechanically and electromechanically uniform.

A substantial increase in the vibration resistance of the hydrophone may be achieved, if to attach the piezoelement to the supporting structure by its middle section, as shown in Figure 3.30 (b). In case that the piezoelement is ideally uniform there should be no voltage at its output in response to vibration in the axial direction due to symmetry. The electrical charges generated in the halves of the cylinder should be equal in value and opposite in sign, as it is seen from the mechanical stress profile shown in the Figure. Thus, $\gamma_{\dot{U}} \approx 0$ and $NI_p^U \to \infty$.

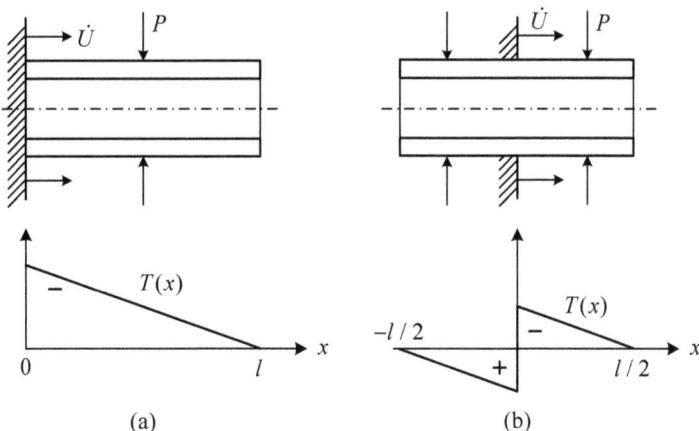

Figure 3.30: Profiles of the stress in a cylindrical piezoelement under action of acceleration depending on the way how it is attached to a supporting structure.

In real situations the halves of a piezoelement are not identical and $\gamma_{\dot{U}} \neq 0$. If it is not small enough for meeting requirements for the noise immunity under particular operating conditions, then the procedure of equalizing sensitivities of the halves of piezoelement to acceleration described above can be used for increasing the acceleration resistance of the hydrophone. Noteworthy is that, if to connect halves of the piezoelement in opposite phase, then transducer becomes ideal (as far as the halves are identical) accelerometer insensitive to sound pressure.

Thus, the noise immunity of the sensors substantially depend on their design, and can be increased by rational designing, as examples in Figure 3.28 and Figure 3.30 show.

Requirement for the admissible value of parameter NI_p^U can be derived from the consideration that the root-mean-square value of the output voltage of a sensor, which is determined due to statistical independence of the signal and noise by the relation

$$V = \sqrt{V_p^2 + (V_p^U)^2} = V_p\sqrt{1+(V_p^U/V_p)^2} \;, \tag{3.209}$$

with a given degree of accuracy must correspond to the voltage V_p generated by the signal only. For fulfilling this condition, it should be $(V_p^U/V_p)^2 \ll 1$. We will assume for estimations that the admissible accuracy is achieved at $(V_p^U/V_p)^2 < 0.1$. Requirement for the parameter NI_p^U under the particular operating conditions depends on relation of the signal P to noise $\dot{U}$ at location of reception. With the assumed admissible value of V_p^U/V_p it is determined by formula

$$NI_p^U \geq 3\dot{U}/P. \tag{3.210}$$

If to apply the similar reasoning to an accelerometer, we will find out that the measure of its noise immunity with respect to the sound pressure is characterized by parameter $NI_{\dot{U}}^P = \gamma_{\dot{U}}/\gamma_{\dot{U}}^P$ (now it is the "sound pressure resistance"). The condition, under which the accelerometer is capable of measuring vibration with the same accuracy as in the previous case, will be

$$NI_{\dot{U}}^P \geq 3P/\dot{U}. \tag{3.211}$$

In essence, only by values of the parameters NI_p^U and $NI_{\dot{U}}^P$ in combination with the known relation $\dot{U}/P$ under particular operating conditions it can be said whether in this situation the sensor is functioning as a hydrophone, or as an accelerometer, or it is unsuitable for measuring either pressure or vibration. In this sense, assigning to sensitivity a subscript that indicates the sensor function is conditional.

For illustration consider as an example a sensor that is intended for measuring vibration of surfaces in air and has for these conditions of operation value of immunity to the sound pressure,

3.2. Transducers in the Receive Mode

$(NI_U^P)_{air}$, sufficiently large for positioning the sensor as an accelerometer. For comparison, also consider the situation where the same sensor is used to measure the vibration of a surface immersed in water. For simplicity of a qualitative estimations suppose that the surface is flat, vibrates in the piston-like mode and is large enough to generate a plane wave. Magnitude of the sound pressure generated will be in this case $P = \dot{U}(\rho c)/\omega$. To measure true value of the acceleration in air the immunity of the sensor to sound pressure must be according to (3.211) $(NI_U^P)_{air} \geq 3(\rho c)_{air}/\omega$. For measuring the acceleration in water with the same accuracy the value of this parameter must be $(NI_U^P)_w \geq 3(\rho c)_w/\omega$. Thus, the immunity to the sound pressure of the accelerometer for measuring in water should be $(NI_U^P)_w = (NI_U^P)_{air} \cdot (\rho c)_w/(\rho c)_{air} \approx 3.5 \cdot 10^3 (NI_U^P)_{air}$, i.e., enormously greater than immunity of accelerometer for measuring in air. Otherwise, the sensor must be considered as a sound pressure hydrophone with noise immunity to acceleration $(NI_P^U)_w = 1/(NI_U^P)_w$.

As we can see, the unwanted actions may cause the serious distortions of the sensor characteristics. Therefore, in the specifications for the sensors the values of their sensitivities to the unwanted actions typical of operating conditions must be presented together with sensitivities to the action of the signal. Note, that among the technical characteristics of the B&K accelerometers the sensitivity γ_U^P to the sound pressure is given along with its sensitivity γ_U to acceleration.

3.3 References

1. S. P. Timoshenko, *Vibration Problems in Engineering*, 2nd Ed. (Van Nostrand, New York, 1937).
2. D. A. Berlincourt, D. R. Curran, and H. Jaffe, "Piezoelectric and Piezomagnetic Materials and their Function in Transducers," in Physical Acoustics, Vol. I, Part A, edited by W. P. Mason (Academic, New York, 1964).
3. P. M. Morse and H. Feshbach, *Methods of Theoretical Physics*, Part I (McGraw-Hill, New York, 1953).
4. P. M. Morse, *Vibration and Sound*, 2nd Ed. (McGraw-Hill, New York, 1948).
5. S. P. Timoshenko, *Strength of Materials*, Part I, Elementary Theory and Problems, 2nd Ed. (Van Nostrand, New York, 1940).
6. E. L. Shenderov, *Wave Problems in Hydroacoustics* (Sudostroyeniye, Leningrad, 1972. (in Russian).
7. L. E. Kinsler, A. R. Frey, A. B. Coppens, and J. V. Sanders, *Fundamentals of Acoustics*, 4th Ed. (John Wiley & Sons, New York, 2000).
8. W. G. Cady, *Piezoelectricity* (McGraw-Hill, New York, 1946).
9. D. Stansfield, *Underwater Electroacoustic Transducers* (Reprinted by Peninsula, Los Altos Hills, CA, 2003).
10. S. P. Timoshenko, *Vibration Problems in Engineering*, 2nd Ed. (Van Nostrand, New York, 1937).
11. L. Ya. Gutin, "A sound field of the piston-like projectors", Zhurnal Tekhnicheskoi Fiziki, Vol. 7, No. 10, 1937. Selected works in Shipbuilding (Sudostroenie, Leningrad, 1977), p. 95 (in Russian).
12. R. S. Woollett, "Procedures of comparing hydrophone noise with minimum water noise," J. Acoust. Soc. Am. **54**(5), 1376-1379 (1973).
13. J. W. Joung, "Optimization of acoustic receiver noise performance," J. Acoust. Soc. Am. **61**(6), 1471-1476 (1977).

LIST OF SYMBOLS

Symbol	Description
A	radius
B	bulk modulus
c, c_c, c_w	sound speed, speed of sound in ceramic composition and in water
c_{mi}^E	elastic stiffness of a piezoceramics at constant electric field
C, C_e^S	capacitance, capacitance of blocked transducer
C, C_{eqv}^E	compliance, equivalent compliance of a mechanical system at constant electric field
d, d_{mi}	separation, distance; piezoelectric constant
D	diameter, flexural rigidity $D = Yh^3/12(1-\sigma^2)$
D_i, D_i^E	charge density, charge density at constant electric field
e_{mi}^E	piezoelectric constant, $e_{mi} = d_{mj}c_{ji}^E$, $j = 1...6$
E, E_{op}, E_p	electric field, operating field, permissible field
Ef	effectiveness
f, f_r, f_{ar}, Δf	frequency, resonance frequency, antiresonance frequency, deviation of frequency
f_{ip}	partial resonance frequencies of a coupled system
F, F_{eqv}	force, equivalent force
G	torsional rigidity
h	height
$H(\theta, \varphi)$	directional factor
I	current
I_L, I_C, I_m	current through inductance, current through capacitance, motional current
J, J_p	moment of inertia, polar moment of inertia
k; k_c, k_{eff}; k_{dif}	wave number $k = \omega/c$; electromechanical coupling coefficient, effective coupling coefficient; diffraction coefficient
k_E, k_T	reserves of the electrical and mechanical strength coefficients
K, K_{eqv}^E, K_{il}	rigidity, equivalent rigidity of a mechanical system, mutual rigidity of coupled systems
ΔK	additional rigidity term that characterizes electrical interaction between elements in nonuniformly deformed piezoelectric body

Symbol	Description
l, t, w	length, thickness, width
L; L_p, L_s	Lagrangian, inductance; parallel and series inductances
ms_w	Mismatch coefficient, $ms_w = r_{ac}/r_{opt}$
ms_i	mode shape coefficient
M; M_{eqv}, M_{il}	Moment, total mass; equivalent mass, mutual mass of coupled systems
n	turns ratio, electromechanical transformation coefficient,
N, N_i	Number of segments in segmented mechanical system, electromechanical transformation coefficients, $i = 1, 3$.
o	subscript that denotes a reference point
P, P_o; P_h	sound pressure, sound pressure of simple source; hydrostatic pressure
Q, Q_e, Q_m	quality factor, $Q = W_{kin}/W_{Loss}$; electrical and mechanical quality factors
r, $\mathbf{r}$	distance, radius vector
r, r_{mL}; r_{ac}, r_{opt}	resistance, resistance of mechanical loss; radiation resistance, optimal value of the radiation resistance
R, R_{eL}	resistance, resistance of electrical loss
s_{mi}^E	elastic compliance of piezoceramics at constant electric field
S, S_{ik}, S_i	deformation, tensor of deformation $(i, k = 1, 2, 3)$, tensor of deformation $(i = 1, .., 6)$
S_Σ, S_{av}, S_{eff}	surface area, average surface area effective surface area
T, T_{ik}, T_i	stress, stress tensor $(i, k = 1, 2, 3)$, stress tensor $(i = 1, .., 6)$
T_{op}, T_p	operating stress, permissible stress
u, U; U_o, U_i	Velocity; velocity of reference point, velocity of reference point in i^{th} mode of vibration
$U_{\tilde{V}}$	volume velocity
v, V	voltage
$\tilde{V}$	volume
w; w_{int}, w_e, w_{mch}, w_{em}	width, energy density; densities of the internal, electrical, mechanical, and electromechanical energies
W, $\dot{W}$, $\bar{\dot{W}}$	energy, energy flux (power), complex power
W_{el}, W_e^S	total electrical energy, electrical energy stored in a blocked piezoelement

List of Symbols

Symbol	Description
W_{int}, W_m, W_{em}, W_{ac}	internal, mechanical, electromechanical, and acoustic energies
W_{kin}, W_{pot}^E	kinetic energy, potential energy at constant electrical field
W_{eL}, W_{mL}	energies of electrical and mechanical loss
$\dot{W}_{mE}$, $\dot{W}_{mT}$	maximum power electric field limited and mechanical stress limited
ΔW	additional energy term that characterizes electrical interaction between elements in nonuniformly deformed piezoelectric body
x; x_{ac}	coordinate; reactance of acoustic radiation
y; $y = \delta/t$	coordinate; ratio of thickness of active layer to total thickness of mechanical system
Y, $Y_i^E = 1/s_{ii}^E$	Young's modulus, Young's modulus of piezoceramics ($i = 1, 3$)
Y_a^E, Y_p	Young's moduli of active and passive materials
Y_σ	$Y_\sigma = Y/(1-\sigma^2)$
z; z_{il}	Coordinate; mutual impedance between modes of vibration
Z, $Z_{il} = z_{il}U_i/U_l$	impedance, introduced impedance
Z_m, Z_m^E, Z_{in}	mechanical impedance, impedance at constant electric field, input impedance
Z_{ac}	radiation impedance
α_{ac}	nondimensional coefficients of the radiation resistance
$\alpha_c = n^2 C_m^E/C_e^S$	coefficient related to effective coupling coefficient, $k_{eff}^2 = \alpha_c/(1+\alpha_c)$
β_{ac}	nondimensional coefficient of the radiation reactance
$\beta = f_{1p}/f_{2p}$	detuning factor between partial frequencies of a coupled system
γ, γ_m, γ_k,	coefficient of coupling between partial systems, coefficients of inertial and elastic coupling
γ_Y	$\gamma_Y = Y_p/Y_a^E$
γ_ρ	$\gamma_\rho = \rho_p/\rho_a$
η; η_{em}, η_{ma}, η_{ea}	efficiency; electromechanical, mechanoacoustic, electroacoustic efficiencies
δ	separation between electrodes,
δ_e, δ_m	angles of dielectric and mechanical losses, $\tan\delta_e = 1/Q_e$, $\tan\delta_m = 1/Q_m$
ε; ε_{ik}^T, ε_{ik}^S	dielectric constant; tensors of dielectric constants of piezoceramics at free and clamped conditions
θ; $\theta(r)$	angle, mode shape

Symbol	Description
λ	wavelength, Lame constant
μ	Lame constant (share modulus)
ξ, ξ_o	displacement, displacement of reference point
ρ, ρ_a, ρ_p	density, density of the active and passive materials
σ, σ_i^E	Poisson's ratio; Poisson's ratio of piezoceramics, $\sigma_1^E = -s_{12}^E / s_{11}^E$, $\sigma_3^E = -s_{13}^E / s_{33}^E$
Σ	surface in general
φ	angle
χ	diffraction function
ω, ω_r, ω_{ar}	angular frequency, resonance and antiresonance frequencies
$\Omega = f^2 / f_{1p}^2$	nondimensional frequency factor
$\Omega = 2\Delta f / f_r$	normalized bandwidth

1. Vectors are displayed in bold letters.
2. Low case letters denoting the time dependent quantities indicate instantaneous values; the capital letters are values in rms.
3. An overbar on a capital letter denotes a complex quantity.

INDEX

A

accelerometer · 69, 70, 74, 144, 148, 149
acoustic interaction · 53, 62, 77, 134
acoustic power · 82, 104, 112, 117, 118, 127
acoustic pressure · 6, 20
acoustic radiation · 19, 104, 122
acoustic wave · 47
admittance · 89, 95, 96
aspect ratio · 53, 135
average surface area · 69, 77

B

baffle · 53, 54, 63, 76
balance of energies · 27
bandwidth · 123
bar transducer · 59, 61, 84, 102, 112, 117, 128
beam · 9, 24, 65, 67, 68, 70
bending moment · 24, 68, 74
Berlincourt · 39, 150
boundary conditions · 10, 33, 65, 70

C

clamped end · 70
clamped transducer · 20, 30
compliance · 12, 13, 49, 50, 151
coordinates · 7, 22, 33, 34, 35
coupling coefficient · 16, 41, 49, 59, 75, 90, 94, 108, 109, 135
cylindrical simple source · 52

D

degree of freedom · 8, 10, 29, 32, 38, 40, 56
diffraction coefficient · 45, 46, 47, 52, 63, 76, 77, 78, 119, 131
diffraction function · 47
directional factor · 5, 54, 70, 81, 103, 131
directivity · 54, 104

E

effective surface area · 67, 73
effectiveness factor · 103
electroacoustic efficiency · 103, 104, 105, 106, 108, 111
electromechanical efficiency · 105
electromechanical energy · 4, 15, 29, 31
energy conservation law · 22, 28, 33
energy flux · 5, 6, 8, 9, 14, 19, 22, 23, 25, 27, 29, 31, 138
equivalent circuit · 31, 32, 83, 131
equivalent generator · 31, 94, 133, 138, 139
equivalent parameters · 36, 56, 59, 61, 68, 69, 70, 71, 74, 77, 90

F

flexure · 24, 66
frequency response · 120, 121, 122, 133, 138
fuler equations · 33

G

generalized coordinates · 6, 7, 33
generalized forces · 6, 7, 35
generalized velocity · 6, 14, 22, 27
generator · 28, 133, 140

H

hydrophones · 65, 75, 129, 131, 133, 137, 138, 146

I

immunity · 130, 144, 145, 146, 147, 148
inertia · 11, 67, 69, 145, 146
intensity · 24, 81, 103, 104, 119
internal noise · 129, 130, 139, 140, 143

K

kinetic energy · 11, 42, 49

L

lagrangian · 33, 34, 35, 36
least action principle · 33

M

matching · 3, 5, 62, 82, 94, 102, 112, 118, 123, 128, 137, 143
mode of operation · 3, 4, 43, 90, 131, 133
mode of vibration · 19, 21, 37, 40, 47, 56, 69, 70, 71, 73, 74, 114, 115

N

noise · 129, 130, 133, 134, 139, 142, 143, 144, 145, 146, 147, 148, 149, 150
noise immunity · 144, 145, 149

O

open circuit · 28, 82, 131, 133, 135
operating conditions · 3, 94, 130, 144, 147, 148, 149
optimization · 128
oscillating disk · 78

P

parallel tuning · 92, 93, 99
permissible electric field · 111, 113, 127
piezoceramics · 15, 18, 57, 62, 67, 109, 134
piezoelement · 16, 17, 40, 41, 62, 67, 68, 73, 90, 91, 109, 114, 136, 137, 144, 146, 147
polarization · 60, 67, 79
potential energy · 12

Q

quality factor · 15, 64, 83, 91, 92, 94, 98, 100, 102, 122, 134, 135

R

radiation impedance · 62, 63
reactance · 91, 95
receive channel · 129, 139
receive mode · 3, 20, 27, 31, 32, 45, 46, 47, 71, 129, 131
reciprocity principle · 46

rectangular beam transducer · 65
reference point · 9, 10, 19, 21, 29, 59, 68, 114
reserves of strength · 118
rigidity · 12, 13, 42, 56
rule of signs · 24

S

self noise · 142
self-impedance · 37
sensor · 139, 140
series tuning · 92, 93, 94, 120
shear · 24, 40
sign convention · 6, 23, 24, 25, 66, 68, 73
signal to noise ratio · 129, 130, 140, 142, 144
simply supported ends · 70
source strength · 45, 52, 77, 78
specific sensitivity · 136, 138, 140, 142, 143
supporting functions · 37

T

tension · 26, 66

thevenin's theorem · 28, 130
Thevenin's theorem · 28
transformation coefficient · 27, 42, 57, 58, 60, 68, 73
transmit channel · 81, 82, 94, 101, 102, 103, 104, 112, 119, 120, 123, 128
transmit mode · 3, 4, 16, 27, 36, 45, 47, 81, 131

U

umov · 22
Umov · 22, 39
underwater · 150
unwanted action · 144

V

volume velocity · 46, 77

W

wavelength · 19, 45, 46, 47, 50, 52, 54, 117

www.ingramcontent.com/pod-product-compliance
Ingram Content Group UK Ltd.
Pitfield, Milton Keynes, MK11 3LW, UK
UKHW051855140426
5217IPUK00006B/127